THE COMPLETE GUIDE TO THE NCO-ER

THE COMPLETE GUIDE TO THE NCO-ER

How to Receive and Write an Excellent Report

Wilson L. Walker
Master Sergeant, U.S. Army, Retired

IMPACT PUBLICATIONS
Manassas Park, VA

THE COMPLETE GUIDE TO THE NCO-ER
How to Receive and Write an Excellent Report

Library of Congress Cataloguing-in-Publication Data

Walker, Wilson L., 1944-
The complete guide to the NCO-ER : how to receive and write an excellent report / Wilson L. Walker
p. cm.
ISBN 0-942710-98-3 : $13.95
1. United States. Army—Non-commissioned officers—Rating of—Handbooks, manuals, etc. 2. United States. Army—Personnel management—Handbooks, manuals, etc. I. Title.
UB323.W277 1993
355.3'38—dc20 93-39157
CIP

For information on distribution or quantity discount rates, Tel. 703/361-7300, Fax 703/335-9486, or write to: Sales Department, IMPACT PUBLICATIONS, 9104-N Manassas Drive, Manassas Park, VA 22111-5211, Distributed to the trade by National Book Network, 4720 Boston Way, Suite A, Lanham, MD 20706, Tel. 301/459-8696 or 800/462-6420.

CONTENTS

ACKNOWLEDGEMENTS

As always, I thank God who makes it all possible, for all of us. I also would like to thank my mother, wife and children.

I have a very special thanks for Tammie whose editing enabled her to translate what I try to say into words that all soldiers can understand and receive the information intended. Also, I would like to thank Impact Publications for turning the information into a book and placing it where it is available to all soldiers.

This acknowledgement would not be complete without thanking all the soldiers for their support in ensuring we all live in a peaceful world. You are the best, and we all wish you success and happiness in the years ahead.

Wilson L. Walker
Master Sergeant, U.S. Army, Retired

This book is dedicated to

my mother's employer,

MR. JAMES L. PERMUTT.

A man that treats all people with the

respect that made him the most

honest and respected man in Alabama.

He is truly the man of all men.

PREFACE

The Non-commissioned Officer Evaluation Report (NCO-ER) is one of the most important reports that you will receive or write as an NCO in the Army. I'm sure you have heard about some NCO's that didn't get promoted, were relieved from duty, or put out of the Army because of bad NCO-ER's.

In this book, I will explain how you can position yourself so that this will not happen to you. As you learn how this is done, you will also be able to write the best NCO-ER's for your NCO's. You will learn about the qualifications and responsibilities of the rater, senior rater, and reviewer. You will learn why bullet comments are so important and which ones are bad for your rating. You will learn what board members and others at DA will look for when they check your NCO-ER and also you will be able to write NCO-ER's for your NCO's just by using this book.

The NCO-ER is so important because it affects the NCO's:

- promotions;
- assignments;
- school selections;
- and retention.

As you can see, the areas affected by your NCO-ER can make your Army career successful or cause you to be put out of the service. The only other thing that has as much influence on your career is the Self Development Test and, because of that, it's important that I tell you something about that also.

Chapter One

TAKING CONTROL

DA FORM 2166-7-1 AND DA FORM 2166-7

When promotion, assignments, schooling or retention time rolls around, members of your branch at the Department of the Army start pulling records. Some of your peers in your MOS may:

- be the same age;
- be the same color or race;
- have the same amount of schooling;
- have the same duty assignment;
- have the same amount of time in grade or service; or
- have the same physical fitness test or other scores.

The chances are rare that your NCO-ER will be the same. What you do is just as important as how your duties are described. As you learn how to take control of your NCO-ER, you can make sure you receive the credit due you. To take charge of your NCO-ER, you need to know about:

- the NCO Counseling Checklist/Record;
- the NCO Evaluation Report;
- your rater;
- your senior rater;
- your reviewer;
- your Commander and First Sergeant; and
- your PSNCO and PAC.

THE NCO COUNSELING CHECKLIST/RECORD

The NCO Counseling Checklist/Record is your NCO-ER guide for counseling and writing the rated NCO's NCO-ER, yet some NCO's may see it one time and there are other NCO's who will never see it. Each time you are counseled or counsel your NCO's, there should be a Checklist/Record along with a working copy of the NCO-ER.

When you are counseled make sure you get a copy of the Checklist/Record and the working copy of the NCO-ER. This is not only for your protection but also to know which direction you are going with your NCO-ER. The purpose of counseling is to improve performance and to professionally develop the rated NCO and, at the same time, improve the rater's counseling skills and writing abilities.

If you find that your rater is not counseling you the way he should, tell him how it should be done. First of all, there has to be a face-to-face performance counseling for all NCO's (Sergeant E-5 to Sergeant Major). In other words, your rater should sit so you are facing each other when he is counseling you about your job performance. The important things he talks to you about during the counseling should be what he wants you to do during the rating period and they should also be written down on the Checklist/Report (DA Form 2166-7-1). If all the raters did this, there would be no reason for a late NCO-ER and the rated NCO would have a good idea as to what kind of report he would receive.

After you get your counseling and have a copy showing what it is that he wants you to do, there is no reason for you to get a bad report unless you did not do what you were told.

During the rating period, you should spend most of your time doing what is on the Checklist/Report. Your first counseling must be conducted within the first 30 days of the rating period and at least quarterly thereafter. Within 30 days after you received your last NCO-ER, you should be counseled for the next one or within 30 days after you get a new rater.

To take control of your NCO-ER, always let your rater know when it is time for your new counseling. He should be telling you, but remind him anyway, because there are preparations he has to make before the counseling. He has to schedule the counseling session, which should be done during duty hours. He also needs to find a place for the counseling so that you and he will not be disturbed.

When you report to your rater for your counseling, he should have with him:

(1) A copy of the last duty description used for your duty position. This should be on the working copy of the NCO-ER. If your duty description was changed after your last counseling, be sure to let him know because what he wants you to do may change also, or what he wants you to do might conflict with your new job description. Always be sure that you are being rated only for the job you are doing.

(2) Your rater should have with him a working copy of the NCO-ER (DA Form 2166-7). Parts 2 and 3, which contain your rating chain and duty description, should be filled out on the working copy of the NCO-ER. Make sure you know who your rater, senior rater and reviewer are and that your duty description is what you do on your day-to-day job.

(3) He should also have with him the names of your new

rating chain members, if there are any. Make sure you keep up-to-date. Know who your rating chain members are, where they work, and their relationship concerning your job.

After your rater has made all the preparations for the counseling, you will report to him for counseling. The first thing he should do is to make sure you and he are alone and will not be disturbed. At the beginning, he may talk to you about something not related to the counseling. This is just so you can feel at ease before the counseling begins.

Once the counseling begins, he will go over your rating chain and let you know if there are any changes. He will then show you the draft duty description on the working copy of the NCO-ER. Read the draft word-for-word and make sure it is what you do day-to-day and not just the job description from AR 611-201. Many NCO's use the job description from the book (AR 611-201) and, because of this, the Department of the Army sees all the NCO's in your MOS doing the same thing. Work with your rater if you have to and draft your duty description trying to fill Block (c) of Part 3 of DA Form 2166-7.

When your rater gets to Part 4 on the working copy, he will explain each value and NCO responsibilities. He will not only tell you what they mean in his view, but also how they relate to your job and what you need to do in each area to receive a success rating. When he explains it all to you, ask him what you need to do in order to receive an excellent rating. As he explains the values and responsibilities to you, he may write in the blank space next to values and responsibilities. This will be his draft for your rating.

After you're counseled and he has gone over the values and responsibilities with you, he will ask you if there are any questions. This is the time for you to clear up anything you are unclear about. Make sure you know what he wants done and how he wants it done. Get it in writing, if possible, and try to get a copy of the report. That's all there is to it.

The next time he counsels you, he will do a follow-up by letting you know what you did and did not do and how well you did it, as well as what he wants done during the next rating period. As your rater explains the report to you, he should ask himself.

- What has happened in response to any discussion you had during the last counseling session?
- What has been done well?
- What could have been done better?

If you have any questions or suggestions, be sure to let him know. There are a few NCO's and officers that feel that as long as they write your NCO-ER, they have some kind of control over you. Don't let that happen to you. Take control of your NCO-ER and you won't have to worry about the leaders with poor leadership skills. Remember your NCO-ER begins with you and the Checklist/Report.

Now let's back up a bit and break this down starting with the rating chain qualifications and responsibilities.

RATING CHAIN QUALIFICATIONS AND RESPONSIBILITIES

It's important to know your rating chain but it's just as important to know their qualifications and responsibilities. Before you start preparing for that Excellent NCO-ER, you need to know the qualifications and responsibilities of the rater, senior rater and reviewer.

There are five areas that you can receive an Excellent, Success or Needs Improvement rating and they are:

- Competence;
- Physical Fitness and Military Bearing;
- Leadership;
- Training; and
- Responsibilities and Accountability.

The Army's goal is for all NCO's to receive a Success rating, which is what most NCO's receive. Your goal should be to receive the Excellent rating which would put you far ahead of your peers. To do this, all you need to do is get at least two Excellent ratings in two of the five areas, ensure all the information in Parts 1 through 3 is correct and complete, receive all "yes's" in Part 4, an "X" in the "Among the Best" block in Part 5 and a "1" in the block for overall performance and overall potential for promotion and/or service in positions of greater responsibility.

With that in mind, let's look at the qualifications and responsibilities of:

THE RATER

Your rater is your first line supervisor. He is the person that you work for and may be an officer, NCO, warrant officer or civilian. Just as you have only one rater, remember you have only one boss. Your rater should be designated on the published rating scheme established by the local commander. Remember, the person that you directly work for is your rater.

There are jobs where members of other U.S. military services may be the Army NCO's rater; however, all qualifications are the same as for the Army raters. Allied forces members cannot be raters. Your rater's primary role is that of evaluation, focusing on performance and performance counseling. He must also do the following things.

- Counsel you on your duty performance and professional development throughout the rating period.
- Counsel you within the first 30 days of each rating period and quarterly thereafter.
- Use DA Form 2166-7-1 along with a working copy of the NCO-ER.
- Prepare a fair, correct report evaluating your duty performance, values/responsibilities and potential.
- Verify Parts I and II of DA Form 2166-7 and enter the PT, height and weight results.
- Date and enter his signature on Part II(a) of the DA Form 2166-7.
- Ensure you understand the organization, its mission, and your role in support of the mission and all of the standards

by which you will be judged.

- Use all opportunities to observe and gather information on your performance.
- Prepare a complete, accurate and fully considered evaluation report which should cover failures as well as achievement.
- Be honest in his evaluation and give you full credit for your achievement and potential.
- Outrank the rated NCO or have more time in grade if the rank is the same.
- Be the first line supervisor of the rated NCO for a minimum of three rated months.

As you can see, your rater is a very important person when it comes to your military career. That's why it is important to know his responsibilities when it comes to your NCO-ER.

Before I move on, let me explain two more things about the rater. As you know, the rater must outrank the NCO that he is rating or have more time in grade. But, what if you were an E-6 section chief and you had an E-6 platoon sergeant that had less time in grade than you? Could he be your rater? Yes, he could if he were on the E-7 promotion list and is serving in an authorized position for the new grade. Another thing the rater must do is maintain the counseling forms (DA Form 2166-7-1) until after the NCO-ER for that period has been approved and submitted to USAEREC. If the rater counseled a corporal, he will keep the corporal's checklist for one year. In the next chapter, I will talk more about the rater as I explain how the forms are to be filled out.

THE SENIOR RATER

Your senior rater must be in your direct line of supervision. In other words, he should be your rater's rater and your reviewer should be your senior rater's rater. Ideally, the rating chain should be constructed so that:

- the rated NCO works for the rater;
- the rater works for the senior rater; and

- the senior rater works for the reviewer.

Your senior rater must be familiar with your performance and potential throughout the rating period. He has to evaluate your potential for promotion and future assignments. Unlike your rater, the senior rater does not have to counsel you but he should monitor your performance evaluation process to include your periodic counseling with the rater. Remember, you will only have face-to-face counseling with your rater.

The senior rater must be senior to your rater by either pay grade or date of rank and he must be designated as the senior rater for a minimum period of two months. If the senior rater is a general officer, he can be the rater and senior rater.

The senior rater's primary role is that of evaluation, focusing on potential, and overseeing the performance evaluation and monitoring. It is the responsibility of the senior rater to obtain the rated NCO's signature. However, this is done most of the time by the first sergeant because the first sergeant assists the senior rater.

The senior rater's responsibilities include the following:

- He will prepare a fair, correct report evaluation of the rated NCO's duty performance, professionalism and potential.
- He will obtain the rated NCO's signature in Part II(c) of DA Form 2166-7.
- He must ensure that the rated NCO is aware that his signature does not constitute agreement or disagreement with the evaluation of the rater and senior rater.

If counseling dates are missing, the senior rater will enter a statement in Part V(e), explaining why the counseling was not accomplished and ensure the APFT and height/weight entries are correct. He must enter the appropriate statement, if the rated NCO refuses or is not available to sign the report, and ensure the statement "Does not meet minimum qualifications" is entered in Part V(e) when the senior rater does not meet the minimum time requirement. The senior rater's box marks are independent of the rater's. There are no specific box mark ratings required of the senior rater based on box marks by the rater.

Let's review what you have learned so far. You need to make sure you:

- are counseled within 30 days of the rating period;
- are counseled every quarter;
- get a copy of the counseling form;
- understand what you are to do and how it should be done;
- know your rating chain; and
- set your goal for an excellent NCO-ER.

THE REVIEWER

The reviewer must be a commissioned officer, warrant officer, command sergeant major or sergeant major in the direct line of supervision and senior in pay grade or date of rank to the senior rater. He is the so-called judge of the NCO-ER. All he has to do is look the report over to see if everything is correct.

If there is a misunderstanding about the NCO-ER, the reviewer will then talk to the rater and/or senior rater to try and clear things up. However, he cannot make the rater or senior rater change their rating. If the reviewer does not agree with the report, he would put his handwritten "X" in the nonconcurrence box in Part II and add an enclosure, not to exceed one page. He will also ensure the rated NCO is provided a copy of the nonconcurrence enclosure.

If the reviewer agrees with the report, then he would check the concur box in Part II. It is also the reviewer's responsibility to ensure that the proper rater and senior rater complete the report. He will also date and enter his name in Part II(d) of the NCO-ER. When the reviewer reviews the NCO-ER, he:

- ensures that the proper rater and senior rater complete the report;

- examines the evaluation rendered by the rater and senior rater; and
- indicates concurrence or nonconcurrence with the rater and/or senior rater.

There is no minimum time requirement for the reviewer.

There you have it—information about the rater, senior rater and the reviewer. The report has other stops to make before it gets to the Department of the Army. These stops are:

- Commander/First Sergeant,
- Personnel Service NCO/PAC, and
- Personnel Service Company (PSC).

Let's examine the roles of each of these stops.

COMMANDER/FIRST SERGEANT

The commander may be a senior rater or reviewer for some of the NCO's in his unit but, mostly, he will be the rater for his officers.

The first sergeant, on the other hand, normally will have no one to rate because the senior NCO will most likely work for one of the officers. As you know, the first sergeant cannot be a reviewer but he is the one that will have to ensure that all the NCO-ER's are done on time and turned in to the PAC or PSC. If there is a change in the unit rating scheme, the first sergeant should notify the NCO's in the unit to include the rater, senior rater, reviewer and the PSNCO. He should then post the new rating scheme on the board. The first sergeant should know when the NCO's ratings are due and the deadlines to have the NCO-ER's completed. If the DA Form 2166-7 is not received from the PSC/PAC, he should contact the PSNCO for its delivery.

If any of the NCO's in his unit change duty assignment, the first sergeant and commander will determine whether or not the NCO's should get a "Change of Rater" NCO-ER. The first sergeant also assists the senior rater in obtaining the NCO's signature in Block (c) of Part II of the report. Once the first sergeant receives the blank NCO-ER's from the PSNCO/PAC, he forwards them to the rating officials for completion and returns them to the PSNCO/PAC when they are done.

The commander and first sergeant work directly with the PSNCO/PAC.

THE PSNCO/PAC

The Personnel Service NCO or Personnel Administration Center's (PSNCO/PAC) job is to monitor the personnel transaction register by unit report (AAC-PO1) with the report (AAC-CO7) and the unit rating schemes to ensure the units are promptly requesting change of rater NCO-ER's when they report position changes, if necessary. They also ensure all NCO's are currently reflected on a rating scheme and receive a NCO-ER whether or not occupying an authorized TDA or TO&E position. By using the unit rating scheme, the PSNCO/PAC determines rating officials, establishes suspense dates for the completion of each NCO-ER, and forwards them to the unit for completion. Upon receiving the completed NCO-ER from the unit, the PSNCO/PAC reviews the NCO-ER to ensure it was completed properly by the rating officials and that the rated NCO signs the report or that an explanation is entered in Part V(e) and forwards the NCO-ER to the Personnel Service Company (PSC).

THE PERSONNEL SERVICE COMPANY (PSC)

When the PSC gets the report, they will complete Part 1 and accomplish necessary records update/change, and submit the necessary SIDPERS transactions. The PSC also makes sure the beginning month on the NCO-ER does not overlap with the ending month covered by the previous report.

They then check Item 35, on DA Form 2-1, and NCO-ER date verified on DA Form 2. The PSC will establish a suspense date to ensure the NCO-ER is completed and forwarded to reach the records section at the Department of the Army no later than 60 days. This is important information because after 60 days you can call the Department of the Army to ensure your NCO-ER is there. The PSC will also provide a copy of the completed report to the rated NCO. When the records section at the Department of the Army gets the NCO-ER, they will also check for errors and file it in the NCO's OMPF.

Now you know the road your NCO-ER takes before and after it gets to you. In the next chapter I'm going to talk about the checklist and working copy of the report so that you will know what should be in each block. By having this information, you can ensure you receive an excellent report and you will also be able to write one for your NCO's that earn it.

Chapter Two

KEYS TO SUCCESS

NCO COUNSELING CHECKLIST/RECORD

The NCO Counseling Checklist/Record (DA Form 2166-7-1) is the foundation of your NCO-ER. It is just as important, if not more important, than your NCO-ER because it all begins there. The Checklist is what your rater will use when he counsels you as to what he wants you to do during the rating periods. During the next counseling, he will look at the Checklist to determine how well you did during the past rating period. Along with the checklist, he will use the "working copy NCO-ER" which I will tell you more about later.

The main reason your rater will be counseling you is to improve your performance. He will do this by telling you how well or poorly you did and, if you didn't do as well as expected, your rater will tell you what you need to do for improvement. He will also counsel you to "professional develop" you so that you will have no worries about your promotions, assignments, schooling and retention. Let me say again, your counseling is very important and it's mandatory for all corporals through command sergeant majors. However, the corporals will not receive an NCO-ER.

When the time comes for your counseling, your rater will not just call you in and begin counseling you because first he has to make preparations.

MAKING PREPARATION

You will be able to tell if your rater made preparation for your counseling because, first of all, he should notify you as to when and where the counseling will be done. That doesn't mean that he will just walk up to you and say, "I will counsel you tomorrow about your NCO-ER." He should give you at least a two or three day notice because there are things you may want to do before the counseling takes place like:

- write down questions and suggestions for the counseling;
- ensure you don't have anything else planned for that time frame;
- double check things you were told to do; and
- get a blank DA Form 2166-7-1, 2166-7 and some carbon paper.

When your rater counsels you, he should always ask if you have any questions or suggestions. If you do, bring them up. Always ask, if nothing else, how he wants you to do things and how well you must do them in order to receive an excellent NCO-ER. If you have any suggestions, bring them up also. He may tell you to have all your soldiers wear the MOPP-4 gear during duty hours next Monday, and you may suggest they wear it during the next field problem so they can get more realistic training.

Remember, "Good leaders are self-motivated, they seek ways to improve their skills and knowledge on their own." When your rater tells you what day and time he will counsel you, double check and make sure you don't have any other commitments. Be sure to take along a blank DA Form 2166-7-1 and 2166-7 just in case he forgot his. Take the carbon paper and have him make a copy for you to keep. The regulations don't call for him to give you a copy but just tell your rater you want to have it to make sure you do what is to be done. Always get a copy if you can, in case your rater loses his or changes his rating because of a one time mistake or because the senior rater or another leader doesn't feel you are doing as good as he says you are. The most important part of the checklist is the counseling record.

THE COUNSELING RECORD

The counseling record is found on Page 2 of DA Form 2166-7-1. There are three parts to the counseling record which are the:

(1) Date of Counseling
(2) Rated NCO's Initials, and
(3) Key Points Made.

There are four blocks under each part.

Under "Date of Counseling", you will find "initial" in the first block and "later" in the other three. In the "initial" block under Date of Counseling, the rater will date this block within 30 days after:

- the rated NCO received his last NCO-ER;
- the soldier was promoted to Sergeant, E-5; or
- the SPC was laterally appointed to Sergeant, E-5.

When the rater puts in the date of the counseling, he must use the six digit system for the year, month and day. After he writes in the date, there should be a total of six digits. If he wants to use the date of 16 Sep 93, he would write 930916. The year is first, then the month followed by the day.

The next three blocks under the date of counseling is titled "Later." Here the rater will write in the dates of the next counseling as he gives them to the NCO. These dates should be three months apart (90 days).

The next block over is the "Rated NCO's Initials" block. This is where you would place your handwritten initials after you are counseled. Remember, never fill this block in until after you get counseled because, when you fill it in, you are saying that you were counseled and that you understand what it is that your rater wants you to do. Placing your initials in the block is the last thing you should do during each counseling. Remember, your initials in no way mean you must do everything you were counseled about. It simply means that you understand what you were counseled about. Never let anyone initial your block for you, even if you are away on TDY or leave. Let me say again, never sign the initial block until after the counseling and you understand what your rater wants you to do.

The "Key Points Made" block is where your rater will write down what it is that he wants you to do. This is the block that most of your questions should be about and this block is the reason you will place your initials in the "Rated NCO's Initials" block. In the "Key Points Made" block the rater should only write down the things that he wants you to pay particular attention to over the next quarter.

The first block under "Key Points Made" next to the "Initial" block is not used to talk about past performance or something you didn't do during the last rating period, only the key point or area should be listed.

It may take more than three months to complete the tasks listed. The rater should list each task by priority and each should be separated with semi-colons. When complete, the block could look something like this:

> Prepare soldiers for AGI; get driving license for all soldiers in section; develop SDT study plan.

The other three blocks under "Key Points Made" next to the "Later" blocks are used to talk about how well the rated NCO did or didn't do the duty or assignment he was given. Also, these blocks are used to describe what should be done during the next rated period and could look something like this:

> Good job on licenses for the soldiers; keep working on SDT study plan; you are making progress for the AGI; do a 100 percent inspection on soldiers TA-50.

When the time comes, the rater will fill in the other "Key Points Made" blocks the same way, first talking about what was or was not done, then what he wants you to do during the next quarter or rating period.

The information your rater puts in the "Key Points Made" blocks will later be transferred to your NCO-ER under "Areas of Special Emphasis" in Part III. The dates of each counseling will also be transferred to your NCO-ER in Part III.

We will return to this point later.

WORKING COPY OF THE NCO-ER

The working copy of the NCO-ER is nothing more than the NCO-ER itself. It could be turned in as your report. Some raters fill in the complete working copy just to let the rated NCO know how he would be rated if he was being rated at the time of the counseling. For this reason, if the rated NCO had to leave the unit, the working copy could be used as the NCO-ER. To keep this from happening, you and the rater should make sure that he writes "working copy" in ink on the top of the DA Form 2166-7 right over "NCO Evaluation Report." As we move on, I will tell you other ways you can make sure your working copy is never used as your NCO-ER.

When your rater starts filling in the working copy of your NCO-ER, the first thing he will do is look over Part I, which is the administrative data.

ADMINISTRATIVE DATA

In Block (a) your rater will write your name; last name first, first name, then your middle initial. Make sure that is the way it is written, not only on the working copy, but the NCO-ER itself. There could be many other NCO's with the same name as yours so double check

Block (b) when he writes in your social security number. Make sure it is correct. Remember, the name may be the same but not the social security number.

The next block, Block (c), is for your rank. Here, your rater would use three letters to show your rank. It could be:

- SGT - for Sergeant;
- SSG - for Staff Sergeant;
- SFC - for Sergeant First Class;
- MSG - for Master Sergeant; or
- SGM - for Sergeant Major.

Block (d) is for your date of rank. Again, your rater will use the six digit system to write in this date. Make sure this is correct. If not, it may keep you out of the primary zone by one day or more.

Block (e), PMOSC, is for your primary MOS code and any special skills you may have. There should be nine digits in this block. There should be at least five for your MOS and the other four should be all zeros unless there are other letters and numbers that relate to your special skill, like the letter "m" that would be used if you are a first sergeant. If you were a 16D, E-8 First Sergeant, the block would be filled in like this: 16D5M0000. Make sure your primary MOS code is correct.

Block (f) is for your unit, organization, station, zip code or APO and major command. It should be listed in that order so if you were in HHB, 3rd BN, 1st ADA, 31st ADA Bde, Fort Hood, Texas 76544 FC, your unit would be Headquarters, Headquarters Battery (HHB), 3rd Bn 1st ADA. The 31st Air Defense Artillery Brigade is the organization and Fort Hood, Texas, is the station with 76544 as the zip code. FC is the code for the major command. You can find the codes for all the major commands in AR 680-29.

Block (g) is the "Reason for Submission" block. Here the rater will put in the two digit code for the type of NCO-ER being submitted. The types of NCO-ER's and codes are:

- 01 - First Report;
- 02 - Annual Report;
- 03 - Change of Rater;
- 04 - Complete the Record; and
- 05 - Relief for Cause.

Block (h) is the "Period Covered" block. The rater does not have to fill in this block on the working copy so have him cross out the blocks so you and he can make sure the working copy would not be used for your NCO-ER. The procedure for computing the time period on the NCO-ER is as follows: Your beginning month is always the month following the ending month of your last NCO-ER. If you receive your annual report on 9308, the beginning month for your next report would be 9309. When you get your first NCO-ER, your beginning date will be the effective date (month) of your promotion to sergeant.

The next block, Block (i), is for rated months. To find the total number of months for this block, all you have to do is add the number of months in Block (h) and put the number in Block (i).

Block (j) is for the non-rated codes. This block, along with Blocks (h) and (i), does not have to be filled out until the NCO-ER is due, not during the counseling session. If there is a code in Block (j), the number of rated months may change depending on the amount of time used for the codes. The codes are:

- A - AWOL or desertion;
- B - Break in service;
- C - Confinement;
- I - TDY or in-transit;
- M - Missing in action;
- P - Patient (including convalescent leave);
- Q - Rater not qualified to rate NCO;
- R - New recruit program;
- S - Military or civilian schooling;
- T - Special Duty;
- W - Prisoner of war; and
- Z - Any reason without a code.

The number of months used for a non-rated code is determined by the number of non-rater days. The time used is as follows:

Total Non-Rated Days	Total Non-Rated Months
15 Days or Less	0 Months
16-45 Days	1 Month
46-75 Days	2 Months
76-105 Days	3 Months
106-135 Days	4 Months
136-165 Days	5 Months

If there isn't a non-rated period, Block (j) will be left blank. Blocks (k), (l), (m), (n), and (o) do not have to be filled in during counseling or when DA Form 2166-7 is used as a working copy.

Block (k) is for enclosures that may be attached to the report should the reviewer check the nonconcur block in Part II(e) of the NCO-ER.

Block (l) will be filled in by PSC using black ink for a handwritten "X" in the "Given to NCO" or "Forwarded to NCO" block.

PSCOIC or PSNCO will initial Block (m) by hand using black ink.

Block (n) is for the major command code which can be found in AR 680-29.

Block (o) is PSC block which is used for their PSC code. Your rater can cross out all the blocks that are not needed for the working copy, so in Part I he can cross out (g), the "Thru" part of (h), and (i) through (o).

AUTHENTICATION

Each time your rater counsels you, he should go over Part II which is mostly your rating chain's "Authentication." If there is a change in the rating chain, he will let you know who will be their replacement. As I said before, try to keep up with this yourself and, if your rater doesn't know about a change, be sure to inform him. When DA Form 2166-7 is used as a working copy, your rater can cross out all block for signatures and dates because there is no need for the working copy to be signed.

When you put your initials on the checklist/record (DA Form 2166-7-1), it shows that your rater has counseled you about what's on the checklist and working copy. Your rater should have all the information filled out in Part II. This is what you should be checking in Part II of the working copy. Block (a) is for your rater's name, social security number, signature, PMOS, branch, organization and duty assignment. The same information will be needed in Blocks (b) through (d) which is for your senior rater and reviewer.

Now, let's back up and look at each block closer. First, there is the name which should be the last name first, the first name next, then the middle initial. Next, is the social security number and block for the signature. This is the same in Blocks (a), (b), and (d). Block (c) is the block for your signature. Information about you is in Part I. When you sign your name in Block (c) of Part II, you are in fact saying that your signature does not constitute agreement or disagreement with the evaluations of the rater and the senior rater. You are also verifying your height/weight and APFT entries in Part I. You are saying that you have seen the complete report (front and back) and that you are aware of the appeals process in AR 623-205. That is the only reason for your signature in Block (c) of Part II.

On the next line in Blocks (a), (b), and (d) is a place for your rater's, senior rater's, and reviewer's rank, PMOSC, branch, organization, and duty assignment. If your rater, senior rater or reviewer is a civilian, he must have senior executive service (SEC) rank and precedence. If he is your senior rater, he must be a G-6 or above, and it takes a G-9 or above to qualify for the reviewer position. Next to the rank, you have a place for the PMOSC/Branch. Here you will find the PMOSC for NCO's and Warrant Officers. The branch will only be used for officers and it will be the branch they were trained for, for example: quartermaster, signal, infantry. After that is the complete organization and duty assignment.

Remember, Block (e) is the for the reviewer who will concur or nonconcur with the report. If you don't agree with the report, you will still sign Block (c) and let the reviewer know you don't agree. Never sign a blank NCO-ER. For your protection, you should not sign a blank report, if:

- You are going on TDY and the report will be done while you are away. Your unit can send it to you.
- You are the rater and must leave the unit for another assignment. Find time to do the report yourself.
- You will be away on leave when the report is due. Make sure it is complete before you leave.
- You are PCS-ing and your report will follow. Do whatever it takes to get your report before you leave. Use the chain of command.

To make a long story short, never, never sign a blank NCO-ER. When your NCO-ER is complete and you are told to sign it, but to not write in a date, that is OK because it depends on the type of NCO-ER as to when it should be signed and also your PSC and PAC have a deadline to meet before they forward it to the Department of the Army. Just make sure Blocks (h), (i) and (j) are correct in Part I of the report. If you are PCS-ing, your NCO-ER can be completed and signed up to 10 days prior to the date of departure in order to facilitate orderly out-processing.

I will go into more detail about the types of NCO-ER's later, for now let's examine Part III.

DUTY DESCRIPTION

This is another part of your rater's job. He has to complete Part III and verify the information with you, the rated NCO. Make sure you go over this part very carefully with the rater. Now, let's break it all down like we did Parts I and II.

First, we have Block (a) which is for your principal duty title and should reflect the actual duty you perform daily. A 24T20's principal duty title is Patriot Operator and System Mechanic (Patriot Op/Sys

Mech), but what if he was the unit mail clerk and that was his only duty besides some odd jobs the commander or first sergeant may have him perform, what would his principal duty title be then? It should be Administrative Specialist. There should have been a Change of Rater NCO-ER done when he was taken from the section to become the mail clerk. If he still works in the section, but is also the mail clerk, then he will have the same rater and the mail clerk job will be an appointed duty, which I will tell you more about later. Remember, the principal duty title is your day-to-day duty. Next, we have Block (b) which is duty MOSC, or your duty MOS code. Your duty MOSC may not be the same as your PMOSC that I told you about for Block (e) in Part I; however, your MOSC should correspond with the principal duty title. To find the PMOSC and MOSC, look in AR 611-1 and update AR 611-201. Block (c) is for your daily duties and scope, to include, as appropriate, people, equipment, facilities and dollar amounts. This block should tell about the most important day-to-day duties and responsibilities. It should also address the number of soldiers you supervise, the equipment, facilities and the amount of dollars involved in your duties and responsibilities. Take the 24T20 I talked about earlier. If his rater looked in the regulation for his duty description, he will find that a 24T20 operates or performs unit maintenance on the Patriot Air Defense Missile system, operates and maintains ICC and ECS.

If your rater is like many of the other raters, he will use the regulations to find your duty description, which means your description will be like many of the other NCO's that have the same MOS and the Department of the Army will look at your report as the same except for the rating. When your rater fills in Block (c), help him by giving him a list of things you do. However, don't try to tell him how to do it. Your list may look something like this:

- Provides technical guidance to lower grade personnel;
- Performs checks, adjustments, and repairs on major item interface;
- Trains operators in operator level maintenance;
- Utilizes proper safety and quality procedures;
- Performs unit maintenance on ICC, ECS, Radar set, IFF,

launching station, anti-radiation missile decoy, and antenna mast group;
- Emplacement of two Patriot launchers; and
- Responsible for three soldiers.

At the same time you give your rater the list, you should also know the value of all your equipment and facilities that you are in charge of. This is very important information and could very well be the reason for your promotion selection at the Department of the Army. Don't forget when your records are being viewed at the Department of the Army that they look at your last five NCO-ER's and they want to see how many soldiers you are in charge of, the dollar amount of the equipment you control, and so on. When your rater gets your list, he can add it to his by using action phrases separated by semicolons, and it could very well look something like this:

> Supervises and performs duties of the Patriot missile launcher section; provides technical guidance to subordinate personnel; trains operators in operator level maintenance; performs maintenance on ICC, ECS, Radar Set, IFF, Launcher station, Anti-radiation Missile Decoy, and antenna mast group; emplaces Patriot missile system launchers; responsible for the health, welfare, morale, and training of three soldiers and over ten million dollars worth of equipment and facilities.

Not only will Block (c) have more to tell about what you do from day-to-day but, most likely, the block will be full, which will make it look more important.

Remember, Block (c) should have the most important day-to-day duties and there should be nothing said about additional duties in this block. "Area of special emphasis" is found in Block (d). You may have or see a NCO-ER with nothing in this block and, when it is checked at the Department of the Army, nothing in Block (d) means that you were not counseled because each time your rater counsels you he will write it down on the checklist. What he writes down in the "Key Points Made" block, he should also put in Block (d) of the working copy and your NCO-ER.

Earlier, I was telling you about the 24T20 that was also the mail clerk. All appointed duties are placed in block (e) and they may not be associated with the duty description. Some raters will put the same thing down for appointed duties, as they do for the principal duty title, so make sure that does not happen to you. Some appointed duties are:

- Mail clerk;
- Commander or First Sergeant driver;
- Gate guard;
- Training NCO;
- Key Control NCO;
- Security NCO;
- Fire and Safety NCO; and
- there are many more.

Most of these duties are appointed to the NCO by the Commander using an additional duty DF. If you have any additional duties, make sure your rater puts them in Block (e).

If you have a very demanding job and two or more additional duties and get a bad report, the additional duties could be what's keeping you from getting a better NCO-ER. On the other hand, if you have two or more appointed duties and get a good NCO-ER it may mean to others that you are a very responsible NCO.

Block (f) is the last block in Part III and it is for your counseling dates from the checklist/record. The same counseling dates your rater put on the counseling records under "date of counseling" is the date he will place on the NCO-ER in Block (f). Absence of the counseling date will not be any help to you should you have to appeal your NCO-ER. Whatever you do, make sure you get your counseling and the date of counseling is put on the counseling records, as well as the NCO-ER working copy and true NCO-ER. If not, your rater, or someone in your NCO-ER rating chain, may just put in some date to make it look like you were counseled.

VALUES/NCO RESPONSIBILITIES

This is the part of the NCO-ER that is reviewed more than any other part because Parts IV and V are where your rater will rate you on your performances and responsibilities. What he puts on the working copy can very well be the final results of the NCO-ER. Part IV is broken down into six parts which are:

- Values
- Competence
- Physical Fitness and Military Bearing
- Leadership
- Training
- Responsibility and Accountability.

A NCO's performance on the Commander's Evaluation (CE), Self Development Test (SDT), Common Task Test (CTT), Army Physical Fitness Test (APFT), Weapons Qualification and compliance with AR 600-9, Army Weight Control Program Standards, must all be considered before completing the evaluation portion of Part IV.

Block (a) under Part IV is all about Values. Values are what soldiers, as a profession, judge to be right. They are the moral, ethical, and professional attributes of character. Values are the heart and soul of a great army.

Part IV(a) of the NCO-ER includes some very important values such as putting the welfare of the nation, the assigned mission, and teamwork before individual interest; exhibiting absolute honesty and courage to stand up for what is right; developing a sense of obligation and support between those who led, those who lead, and those who serve alongside; maintaining high standards of personal conduct on and off duty; and, finally, demonstrating obedience, total adherence to the spirit and letter of a lawful order, discipline, and ability to overcome fear despite difficulty or danger.

When your rater rates you in Part IV(a) about your values, there are seven questions he can answer either yes or no. If he should answer no to any of the seven, he must make a comment as to why. There is no comment needed for a yes answer, but it would make the report more attractive. There is space below the seven questions for the rater to make his bullet comments. Remember, a comment is mandatory for a no answer but not for a yes answer.

Bullet comments are so important on your NCO-ER that the entire next chapter will be devoted just to talking about why they are and how to receive good ones. The seven questions or statements your rater will answer yes or no to in Block (a) of Part IV are:

1 Places dedication and commitment to the goals and mission of the Army and nation above personal welfare.
2. Is committed to and shows a sense of pride in the unit—works as a member of the team.
3. Is disciplined and obedient to the spirit and letter of a lawful order.
4. Is honest and truthful in word and deed.
5. Maintains high standards of personal conduct on and off duty.
6. Has the courage of convictions and ability to overcome fear—stands up for and does what's right.
7. Supports EO/EEO.

On the working copy, the rater should write yes or no to the statement in pencil. Should you get a "no" rating for any of the seven statements, your rater should counsel you about it on the counseling checklist (DA Form 2166-7-1). Most NCO's get an "X" in the yes block for all seven statements. Number 7, "Supports EO/EEO", is the one statement that you do not want to get a "no" because if you do, you can be put out of the Army. We have now completed the front side of DA Form 2166-7.

Turning the form over on the back side, you will find at the very top, three blocks, one for the rated NCO's name, one for his social security number, and one for the "Thru" date. All this information should be the same as on the front side to include the "Thru" date. However, the "Thru" date will not be needed for the working copy.

EXCELLENT, SUCCESS, AND NEEDS IMPROVEMENT RATINGS

If you get an Excellent rating, this means you exceed the standards set for the area in which you received the Excellent rating. The Excellent rating is demonstrated by specific examples and measurable results, both special and unusual. It is achieved by only a few who are clearly better than most others. Try to get at least two Excellent ratings in Part IV of your NCO-ER. This will put you ahead of most of your peers.

Success ratings mean that the NCO was successful in doing what he was told to do. The Army's goal is to have all NCO's get a Success rating. Your goal should be to receive the Excellent ratings which will put you ahead of those that only want to survive.

The Needs Improvement rating is the one that can get you put out of the Army. It doesn't matter if you need some or much improvement. You should never get a Needs Improvement rating without having been counseled about what it was that you did not do. If a NCO received one or more Needs Improvement rating in Part IV (b)through (f), he cannot receive a rating of "Among the Best" in Part V(a). Now, let's see how the ratings are looked at for promotion.

- An Excellent rating means promote the NCO now.
- A Success rating means promote the NCO if there are any stripes left after all the NCO's with Excellent ratings have been promoted.
- The Needs Improvement rating means do no promote, try QMP.

The rater will indicate the rated NCO's level of performance for each responsibility by placing his handwritten "X" in the appropriate box.

The rater must remember that evaluation comments, favorable or unfavorable, shall not be based solely on a noncommissioned officer's marital status. They shall not be made about the employment, educational, or volunteer activities of a noncommissioned officer's spouse.

The rater should also consider and use the nine leadership competencies in FM 22-100, Military Leadership, with the appropriate NCO values and responsibilities (Pages 3 and 4 of DA Form 2166-7-1 and Part IV of DA Form 2166-7) when conducting performance counseling sessions. The leadership competencies are: communications; supervision; teaching and counseling; soldier-team development; technical and tactical proficiency; decision making; planning; use of available systems; and professional ethics. Now, it's time to get back to the areas in which you will be rated.

Part IV(b) is Competence. Under the word Competence is a list of five statements that are associated with competence. These statements will give the rater an idea of what he should be looking for when he rates the NCO in this area. Competence is the knowledge, skills, and abilities necessary to be expert in the current duty assignment and to perform adequately in other assignments within the MOS when required. Competence is both technical and tactical and includes reading, writing, speaking and basic mathematics. It also includes sound judgement and the ability to weigh alternatives, form objective opinions and make good decisions.

Closely allied with competence is the constant desire to be better, to listen and learn more, and to do each task completely to the best and achieve them, create and innovate, take prudent risks, never settle for less than the best, and be committed to excellence. Here are the statements that are listed under competence in Part IV(b).

- Duty proficiency; MOS competency
- Technical & tactical: knowledge, skills and abilities
- Sound judgement
- Seeking self-improvement; always learning
- Accomplishing tasks to the fullest capacity; committed to excellence

Looking at the statements under competence, you can see that it will not be hard at all to get an Excellent rating in this area. Getting 100 percent on the Self Development Test should justify the Excellent rating because it covers duty proficiency, MOS competence, technical and tactical, skills, knowledge and abilities, seeking self-improvement, always learning and committed to excellence. Again, talk to the rater and find out what you need to do in this area to get the rating you desire. For this and all other areas, if you can't get the Excellent rating, shoot for the Success rating with three bullet comments.

In Part IV(c), we have Physical Fitness and Military Bearing. Physical fitness is the physical and mental ability to accomplish the mission-combat readiness. Total fitness includes weight control, diet and nutrition, smoking stoppage, control of substance abuse, stress management, and physical training. It covers strength, endurance, stamina, flexibility, speed, agility, coordination, and balance. Military bearing consists of posture, dress, overall appearance, and manner of physical movement. Bearing also includes an outward display of inner feelings, fears and overall confidence and enthusiasm. It also includes on-the-spot corrections. The statements that are listed under physical fitness and military bearing in Part IV(c) are:

- mental and physical toughness;
- endurance and stamina to go the distance; and
- displaying confidence and enthusiasm, looks like a soldier.

All NCO's should be able to get an Excellent rating in this area because there are so many ways you can do it. First of all, the APFT numerical scores will be entered to justify an Excellent or Needs Improvement rating based solely on the APFT. I remember when I got an Excellent in Physical Fitness and Military Bearing because I made 300 on my last APFT. The bullet comment read "made 300 on last PT test." Maybe you don't need to make 300 to get an Excellent rating because the Army standard for Excellent in physical fitness is 290 on the test, that's why you are given the Army Physical Fitness badge "patch" when you make 290 or above on the PT test. Also, commanders are encouraged to commend soldiers that score over 270 on the APFT.

There are other things that you can do to get an Excellent for Physical Fitness and Military Bearing, like talking to your first sergeant about being in charge of all the overweight soldiers and helping them lose weight. You could help soldiers to stop smoking, set up some stress management classes or help soldiers improve their APFT scores. You could also pick up the Excellent rating just by having outstanding military bearing. Here is a list of regulations that can help you get that rating you want in physical fitness and military bearing.

- AR 350-15 - The Army Physical Fitness Program
- AR 215-1 - The Army Sports Program
- AR 600-9 - The Army Weight Control Program
- AR 600-85 - The Army Alcohol and Drug Abuse Program
- AR 40-25 - Nutritional Standards
- AR 600-63 - The Army Health Promotion Program
- AR 635-200 - Fail APFT "Enlisted Soldiers"
- AR 635-100 - Fail APFT "Officers"
- DA Pam 350-22 - Help improve APFT scores
- AR 600-8-2 - Flagged for failing the APFT
- FM 21-20 - Physical Fitness Training

In Part IV, on the right-hand side of Block (c), are two small blocks, one for the APFT date and one for whether the NCO passed or failed the test; also, if there is a profile, the year and month the profile was awarded. If the NCO passed the test, the year and month the test was taken (using four digits) will be placed in the block along with the word "pass." If the NCO failed the test, the year and month the test was taken would be placed in the block along with the word "fail." If the NCO had a profile, the year and month profile was awarded would be placed in the block along with the word "profile." These entries will reflect the status of the most recent APFT within the 12 months prior to the last rated days of supervision.

The next small block for the height and weight of the rated NCO will show the height in inches and weight in pounds and an entry of "yes" or "no" to indicate compliance or noncompliance with the provision of AR 600-9 (weight control). There is no date for this block but the date should be the same as the date of the APFT because whenever the test is given a weigh-in is given at the same time. Bullet comments are mandatory for Block (c) to explain:

- the absence of the height and weight data;
- any entry of "no" indicating noncompliance with AR 600-9;
- the basis for a "yes" entry when an individual exceeds the

weight for height screening table limit but, through a body fat determination, is in compliance with body fat standards IAW AR 600-9;

- pregnant NCO's height and weight data left blank during the period of pregnancy. The rater must enter the following statement in Part IVc, "exempt from weight control standards of AR 600-9."

Again, if you can't get an Excellent rating, shoot for the Success rating with three bullet comments.

Leadership is covered in Block (d) of Part IV. If you were to get only one Excellent rating, try to get it in this block. Leadership is the process of influencing others to accomplish the mission by providing purpose, direction, and motivation. The led, the leader, the situation and communication are the four major factor's of leadership. The leader (NCO) must set tough, but achievable, standards and demand that they are met. Caring deeply and sincerely for subordinates and their families and welcoming the opportunity to serve them, conducting counseling (FM 22-101), and setting the example by word and act/deed, can be summarized by:

- BE—Beliefs, values and norms.
- KNOW—The four factors of leadership: understand standards, yourself, human nature, your job and your unit mission.
- DO—Provide purpose, direction and motivation. Instill the spirit to achieve and win. Inspire and develop excellence.

The statements listed under the word "Leadership" in Block (d) are:

- mission First;
- genuine concern for soldiers;
- instill the spirit to achieve and win; and
- Set the example: BE, KNOW, DO.

Just looking at the statements you can see that all you need to do in order to get an Excellent rating in leadership is to take care of your soldier and your mission. Just be a good leader. Set the example. Don't say, "Do as I say," but tell your soldier to, "Do as I do." You can set the example by:

- making 270 or more on the APFT;
- scoring high on the SDT;
- becoming an expert with your weapon; and
- standing up for what is right.

Leadership is not hard at all if you care about your soldiers.

Now it is time to look at training, which is Block (e) of Part IV. Training is preparing individuals, units, and combined arms teams for duty performance, the teaching of skills and knowledge. NCO's contribute to team training and are often responsible for unit training (squads, crews, sections), but individual training is the most important, exclusive responsibility of the NCO corp. Quality training bonds units, leads directly to good discipline, concentrates on wartime mission, is tough and demanding without being reckless, is performance oriented, sticks to Army doctrine to standardize what is taught to fight, survive, and win, as small units when air/land battle action dictates.

Good training means learning from mistakes and allowing plenty of room for professional growth. Sharing knowledge and experience is the greatest legacy one can leave subordinates. Remember, the Army's philosophy is that a NCO should be trained and possess the skills, knowledge and attitudes necessary for successful leadership before assuming the duties and responsibilities of the next higher grade. Not only that, but he also must be able to train others to do the same. The statements found under training Block (e) are:

- individual and team;
- mission focused; performance oriented;
- teaching soldiers how; common task duty-related skills;
- sharing knowledge and experience to fight, survive and win.

To get an Excellent rating in training, all you need to do is train your soldiers as a team. Train them to be the best when it comes to CTT, NTC, PT, weapons and, if you want to go all out, come up with a way to train them to get high scores on the Self Development Test. Always find time to train your soldiers. At no time should they be sitting around doing nothing. Once you are known as a successful trainer, you will become known as a leader.

There's only one more block to cover in Part IV and that is Block (f), which is Responsibility and Accountability. This block covers the proper care, maintenance, use, handling, and conservation of personnel, equipment, supplies, property, and funds. Maintenance of weapons, vehicles, equipment, conservation of supplies, and funds are a special NCO responsibility because they are linked to the success of all missions, especially those on the battlefield. This responsibility includes inspecting soldier's equipment often, using manuals or checklists; holding soldiers responsible for repairs and losses; learning how to use and maintain all the equipment soldiers use; being among the first to operate new equipment; keeping up-to-date component lists; setting aside time for inventories; and knowing the readiness status of all weapons,vehicles, and other equipment. It also includes knowing where each soldier is during duty hours; why he is going on sick call; where he lives, and his family situation. It involves reducing accidental manpower and monetary losses by providing a safe and healthful environment; creating a climate which encourages young soldiers to learn and grow, and to report serious problems without fear of repercussions.

Also, the NCO must accept responsibility for his actions as well as those of his soldiers. The statements listed under Responsibility and Accountability are:

- care and maintenance of equipment and facilities;
- soldiers and equipment safety;
- conservation of supplies and funds;
- encouraging soldiers to learn and grow
- responsible for good, bad, right, and wrong.

Now, as you can see, this is another area in which it is not hard to get an Excellent rating. Make sure you and your soldiers have the best record for maintenance of the equipment during AGI's, CI's, and other inspections. Soldiers and equipment safety works hand-in-hand, prepare for it before field training and during day-to-day duties. Train and encourage your soldiers to learn.

The Self Development Test is for NCO's but there is nothing in the books that says you cannot start training E-4's and below for it. Make sure all your soldiers have a high school education or GED. Prepare them for promotion by setting up a pre-board. Stand up for what is good, bad, right, or wrong. Use the chain of command. If these things don't work, try the EO or the post IG. Make sure you are right and let everyone know you will stand by your decision whatever the outcome may be.

So, there you have it. Now all you have to do is make plans for getting the rating that you want.

The rater has two other blocks in Part V to rate the NCO. It's important to know about these last two ratings.

OVERALL PERFORMANCE AND POTENTIAL

There are only five blocks for ratings in Part V, and two of them are for the rater. Many raters and senior raters think Part V is the for senior rater, which is not true. However, many senior raters will go along with the rater's rating and align their rating with his. At the same time, there are some senior raters that don't think the rater is doing a good job and, for that reason, may rate the NCO better than the rater. The senior rater's box marks are independent of the raters. There are no specific box mark ratings required of the senior rater based on box marks by the rater.

Part V is structured for the NCO'S potential for overall performance and potential consists of, and includes, rater box marks for promotion and service potential; rater's specific position recommendation; senior rater's overall performance and potential; and senior rater's choice of alternative for future performance.

When the senior rater does not meet minimum time requirements for evaluation for rating the NCO, the statement "Does not meet minimum qualifications" will be placed in Block (e) in Part V of the

NCO-ER. The senior rater's bullet comments should focus on potential, but may address performance and/or evaluation rendered by the rater. He must also comment on marginal, fair, or poor ratings in Part V.

The raters has only two blocks for ratings in Part V. The first is Block (a) which is the rated NCO's overall potential for promotion and/or service in positions of greater responsibility. The rater can place his handwritten "X" in one of three boxes, which are "Among the Best," "Fully Capable," and "Marginal." Because most NCO's get all Success ratings, the rater may place an "X" in the "Fully Capable" box just because it is in-line with the Success rating. The "Among the Best" rating is the best of the three. However, if a NCO receives one or more "Needs Improvement" ratings in Part IV (b) through (f), he cannot receive a rating of "Among the Best."

The "Among the Best" rating means the NCO has demonstrated a very good, solid performance and there is a strong recommendation for promotion and/or service in positions of greater responsibility.

The "Fully Capable" rating means the NCO has demonstrated a good performance and, should sufficient promotion allocations be available, the NCO should be promoted.

The "Marginal" rating means that the NCO has demonstrated poor performance and should not be promoted at this time.

As you can see, you can get all Success ratings and still get an "Among the Best" rating in Block (a) of Part V. Three Bullet comments in Blocks (b) through (f) of Part IV can also be helpful in getting the "Among the Best" rating in Part V.

The next rater's block in Part V is Block (b). Here, the rater can list three positions in which the rated NCO could best serve the Army at his current or next higher grade. At least two duty positions will be entered, but no more than three. Should the NCO be rated "Among the Best", Block (b) should show two or three areas he could best serve the Army in his next grade because an "Among the Best" rating means the NCO should be promoted. At the same time, if the NCO should receive a "Marginal" rating, the two or three positions he could best serve the Army may show the one he is now serving plus the one below the position he is now in. That takes care of the two rater's block. Now I will tell you about the three senior rater's blocks.

The first senior rater's block is Block (c) in which he will rate the NCO on his overall performance. In order to do that, the senior rater can rate the NCO as successful with a number 1, 2, or 3 or he could rate the NCO fair or poor. If the senior rater rates the NCO successful with a number 1, it means that the NCO is a very good performer and is a strong recommendation for promotion. It is about the same as a rater's "Among the Best" rating. If the senior rater rates the NCO successful with a number 2, it means the rating is about the same as the number 1, but not as strong. A successful rating with a number 3 means the NCO is a good performer and should be promoted if sufficient promotion allocations are available. That rating is about the same as the rater's "Fully Capable" rating. A fair rating by the senior rater means the NCO may require additional training/observation and should not be promoted at this time, which is the same as the rater's "Marginal" rating. The worst rating a NCO can receive from the senior rater and, one that can cause him to be put out of the Army, is the poor rating. A rating of poor means that the NCO is weak or deficient and, in the opinion of the senior rater, the NCO needs significant improvement or training in one or more areas. It also means "Do not promote."

The next block for the senior rater is Block (d) which is for the rated NCO's overall potential for promotion and/or service in positions of greater responsibility. The ratings in this block have the same meaning as Block (c), only he will rate the NCO superior with the number 1, 2, or 3 instead of successful with the number 1, 2, or 3.

The next, and last block, is for the senior rater's bullet comments. In this block, he can write up to five comments. Remember, he must comment on marginal, fair or poor ratings in Part V of the NCO-ER (DA Form 2166-7).

Chapter Three

BULLET COMMENTS

A bullet comment is a statement that may or may not have a verb, object, or subject. It is a short, concise comment used by raters to justify their evaluations. Because they require the raters to make specific points, bullet comments are hard to inflate. The ideal bullet comment is only one line, but no more than two and no more than one bullet per line. Each bullet comment must be double-spaced. There can be no more than three bullet comments in Blocks (a), (b), (c), (d), (e) and (f) of Part IV and no more than five bullet comments in Block (e) of Part V of DA Form 2166-7. Each bullet comment will be preceded by a small "o" to designate the start of the comment. The best bullet comment starts with action words (verbs) or possessive pronouns (his/her). Avoid using the NCO's name or the personal pronouns he and she. No bullet comment can be used more than once on the NCO-ER.

There must be a bullet comment for all Excellent and Need Improvement ratings, however, the rater does not have to write a bullet comment for the Success ratings. Because of this, many raters select the Success rating box for their rating. If you get a Success rating in leadership or another area with no bullet comment and one of your peers gets the same rating with the bullet comment, who do you think members at DA would think is the best NCO? Many NCO's get a Success rating with one comment, some get a Success rating with two comments, but not many get a Success rating with three bullet comments, and that is why I say try to get an Excellent rating or Success rating with three bullet comments.

I will give you a list of bullet comments for each area in Parts IV and V so you will have an idea as to the kind of bullet comment you may want to write. The bullet comments in this book are only examples. Using the following word list, you can add to or take away to form your own bullet comments.

EXCELLENT AND SUCCESS ADJECTIVES AND VERBS

Accurate
Ace
Active
Affirmative
All-around
Alert
Appealing
Calm
Candid
Capable
Charismatic
Clear-thinking
Competent
Complete
Composed
Concise
Confident
Consistent
Contractive
Cooperative
Courageous
Courteous
Creative
Curious
Decisive
Dedicated
Dependable
Determined
Diligent
Diplomatic
Discreet
Dynamic
Eager
Effective
Efficient
Eminent
Enthusiastic
Excellent
Exceptional
Expert
Extraordinary
Extreme
Fair-minded
Favorable
Fearless
Fine
First-string
Flexible
Forceful
Foremost
Forward-looking
Gallant
Generous
Genuine
Good-humored
Good-natured
Gung-ho
Helpful
High
Honest
Imaginative
Important
Independent

Innovative
Intense
Involved
Knowledgeable
Loyal
Major
Mature
Maximum
Meaningful
Motivated
Neat
Objective
Open-minded
Optimistic
Orderly
Organized
Original
Outgoing
Outstanding
Patient
Perceptive
Perfect
Persevering
Persuasive
Pleasant
Polished
Positive
Powerful
Practical
Precise
Predictable
Productive
Professional
Progressive
Proper
Punctual
Quick
Rational
Realistic
Remarkable
Resourceful
Respectful
Responsive
Self-confident
Self-demanding
Significant
Sincere
Sizable
Sound
Special
Splendid
Stern
String
Successful
Superior
Supportive
Systematic
Tactful
Thorough
Trustworthy
Understanding
Unique
Valuable
Vigorous
Well-liked

There are a countless number of other adjectives you could find and use but to put them all in this book would be a book in itself. As you find more, you can put them in this book for your use. Now, for the list of verbs.

Accepts
Achieves
Accomplishes
Accounts
Acquaints
Acquires
Acts
Actuates
Adapts
Adheres
Adjusts
Administers
Advances
Advises
Advocates
Agitates
Analyzes
Anticipates
Applauds
Applies
Appraises
Appropriates
Approves
Arises
Arouses
Arranges
Articulates

Ascends
Aspires
Assembles
Asserts
Assigns
Assists
Assumes
Attains
Attempts
Authorizes
Avoids
Bolsters
Builds
Capitalizes
Calculates
Carries out
Challenges
Checks
Circumvents
Collaborates
Commands
Communicates
Complies
Comprehends
Computes
Conceives
Concentrates
Conducts
Conforms
Confronts
Considers
Consolidates
Consults
Contemplates
Continues
Contributes
Controls
Conveys
Cooperates
Coordinates
Copes
Creates
Cultivates
Dedicates
Delegates
Demonstrates
Deters
Determines
Develops
Devises
Devotes
Directs
Discusses
Displays
Distinguishes
Dominates
Drafts
Effects
Embodies
Emerges
Emphasizes
Employs
Emulates
Encourages
Endeavors
Energizes
Enforces
Enhances
Enlightens
Enriches
Entices
Erupts
Establishes
Evaluates
Evidences
Examines
Exceeds
Excels
Excites
Executes
Exercises
Expands
Expects
Expedites
Exploits
Explores
Expresses
Fabricates
Faces
Facilitates
Fine-tunes
Focuses
Follows-up
Formulates
Fortifies
Fulfills
Gains
Generates
Gives
Glorifies
Grasps
Handles
Hastens
Helps
Identifies
Ignites
Illustrates
Immerses
Implements
Imposes
Impresses
Improves
Improvises
Influences
Informs
Initiates
Inspects

Inspires
Instigates
Instills
Insures
Interacts
Interprets
Interviews
Issues
Judges
Keeps
Knows
Launches
Learns
Maintains
Makes
Manages
Manipulates
Meets
Motivates
Negotiates
Notifies
Nourishes
Obtains
Operates
Organizes
Originates
Outlasts
Overcomes
Oversees
Overwhelms
Paces
Participates
Perceives
Performs
Persists
Persuades
Plans
Possesses
Practices
Prepares
Prods
Presumes
Prevents
Processes
Produces
Projects
Promotes
Proposes
Prospers
Provokes
Purges
Pursues
Quantifies
Quickens
Radiates
Rallies
Realizes
Receives
Recognizes
Recommends
Records
Reflects
Regards
Regulates
Reinforces
Relates
Releases
Relies
Renew
Reorganize
Reports
Represents
Requires
Respects
Responds
Reviews
Revives
Revises
Schedules
Secures
Seeks
Serves
Shows
Solves
Sparks
Spearheads
Stimulates
Strengthens
Strives
Studies
Supervises
Supports
Surpasses
Sustains
Takes
Thrives
Transforms
Tolerates
Trains
Treats
Understands
Uses
Utilizes
Verifies
Weighs

There you have it. A list of words you can use to write Excellent and Success bullet comments. It is not hard to write a good NCO-ER. The bad ones are the hard ones to write because there are times you may want to show the NCO needs more training but, at the same time, you don't want him to be forced out of the Army. The following list of words will help you write a report for the NCO that is not doing as well as you think he should.

NEED IMPROVEMENT NOUNS, VERBS, AND ADJECTIVES

Adversity
Blemish
Chagrin
Confusion
Defect
Demerit
Demise
Dependency
Detriment
Deviate
Disadvantage
Disappointment
Disaster
Discord
Discredit
Disorder
Dissatisfaction
Error
Excuse
Eyesore
Failure
Fault
Fizzle
Flaw
Forfeit
Friction
Hardship
Harm
Hindrance
Imbalance
Impurity
Imperfection
Impossibility
Inaction
Inadequacy
Inattention
Inconsistency
Inconvenience
Inefficacy
Inefficiency
Infraction
Insignificance
Interference
Intrusion
Invalidity
Irregularity
Lack
Lag
Lapse
Liability
Loser
Misfortune
Mix-up
Nonsense
Nuisance
Opposition
Oversight
Pitfall
Problem
Regress
Regression
Restraint
Rigor
Shortcoming
Shortfall
Sloppy work
Tension
Uncertainty

There are some very harsh words you can use to write a bad NCO-ER so be very careful about the words you select. Remember, the NCO should always be counseled if he is not performing as you think he should. Now, for a list of verbs that can be used for the "Needs Improvement" rating.

Collapse
Conceal
Concede
Condescend
Crime
Debase
Denounce
Deprive
Destroy
Deteriorate
Dilute
Diminish
Disappoint
Disrupt
Dodge
Dwindle
Elude
Encumber
Erode
Eschew
Exaggerate
Extenuate
Fade
Fail
Fake
Falter
Flop
Flounder
Flunk
Foil
Gloom
Hamper
Harm
Hinder
Immobilize
Impair
Impede
Incapacitate
Irritate
Lack
Lag
Lapse
Lavish
Lessen
Limp
Lower
Mistake
Negate
Neglect
Obstruct
Oust
Pall
Quit
Rebut
Refuse
Reject
Relinquish
Restrain
Retard
Stagnate
Suppress
Sway
Transgress
Violates
Vitiate
Weaken
Wear-out
Wilt
Work-over
Worsen

Here's a list of adjectives that can be used to write a bad or not so good NCO-ER. Again, be careful of the words you choose.

Abnormal
Adverse
Amiss
Conflicting
Cumbersome
Defective
Deficient
Desperate
Disappointing
Dispassionate
Disruptive
Detrimental
Effortless
Elusive
Erosive
Erratic
Evasive
Erroneous
False
Faulty
Fearful
Feeble
Forbidden
Formless
Fragile
Fruitless
Gloomy
Good-for-nothing
Grave
Grievous
Grim
Gross
Haphazard

Hard-put
Harsh
Helpless
Hit-or-miss
Hopeless
Humiliating
Idle
Ill-advised
Illegal
Ill-fated
Ill-gotten
Illicit
Imaginative
Imperfect
Impossible
Impotent
Impractical
Imprecise
Improbable
Improper
Inaccurate
Inactive
Inadequate
Inadvisable
Inappropriate
Incapable
Incomparable
Incompatible
Incomplete
Incomprehensible
Inconclusive
Inconsistent
Inconsiderable
Inconvenient
Incorrect
Indefensible
Ineffective
Inefficient
Ineligible
Inert
Inexact
Inexcusable
Inferior
Inopportune
Inordinate
Insecure
Insignificant
Insoluble
Insubstantial
Insufferable
Insufficient
Insupportable
Intolerable
Intolerant
Intractable
Intrusive
Invalid
Irregular
Irrelative
Irrelevant
Last
Last-ditch
Limited
Lost
Low-grade
Low-level
Ludicrous
Meaningless
Miserable
Negative
Negligent
Nonproductive
Null
Obscure
Obsolete
Outcast
Outlandish
Out-of-date
Outrageous
Overdue
Pathetic
Petty
Problem
Purposeless
Redundant
Resistless
Rigorous
Rough
Run-down
Rusty
Scant
Shabby
Skeptical
Skepticism
Sloppy
Small-scale
Somber
Sorrowful
Sparse
Spotty
Stagnant
Subnormal
Substandard
Thriftless
Tiresome
Tricky
Uncertain
Uneasy
Unfavorable
Valueless
Wanting
Washed-up
Wasteful
Weak
Weariful
Wearisome

Now you have a small list of words that you can use to write your bullet comments. Make sure your comments justify the rating. If you should receive a NCO-ER and feel that the bullet comments do not justify the rating, let your rater or senior rater know. At the same time, you can write down two or three that you may want him to use. The best way to do this is to write down two or three bullet comments that you would like to get on your report, then write down two more that you know are more than should be said about your rating. Most likely, the rater will pick the same ones you wanted.

Now, I am going to give you a list of bullet comments that might help you relate to the one you are writing. You will find comments that you can use in all rated areas, which are:

- Values;
- Competence;
- Physical Fitness and Military Bearing;
- Leadership;
- Training;
- Responsibility and Accountability;
- Overall Performance; and
- Overall Potential for Promotion.

Under each heading, I will give you a list of bullet comments that can be used for the Excellent, Success or Needs Improvement rating. You should be able to tell which comment is for what rating. Remember, the Excellent rating is for something that few NCO's accomplish. The Success rating is something that most NCO's receive, and the Need Improvement rating is for the NCO that is not doing something right. Let's get started by examining different types of bullet comments.

BULLET COMMENTS FOR VALUES

POSITIVE VALUES STATEMENTS

- supports military and civilian chain of command
- places unit needs and goals first
- does what is needed without being told

- accepts full responsibility of soldiers and self
- is willing to sacrifice
- honest, upright, sincere and candid
- avoids deceptive behavior
- does the right and moral thing
- is a good role model
- develops self and soldiers
- soldiers ethically
- shares thought process with soldiers
- chooses action that best serves the nation
- unparalleled loyalty
- is loyal to organization
- displays absolute loyalty to superiors
- builds loyalty in soldiers
- places unit's interest ahead of own
- is committed to unit goals
- displayed loyalty during "Operation Desert Storm"
- extremely dedicated to the cause
- takes pride in job performance
- displays loyalty in profession
- effectively handles pressure
- effectively manages stress
- makes positive use of stress
- performs well under pressure
- performs well under stress
- performed well during "Operation Desert Storm"
- works effectively in high pressure situations
- gets the job done
- gains control over job pressures
- great personal drive
- keen sense of ethical conduct
- admirable courage
- never gives up on mission
- honest and faithful
- stands behind principles
- proper personal behavior
- high ethical principles
- great personal behavior
- shows courage under pressure

- composed under pressure
- engaging personality
- never loses temper
- sound character
- friendly and sociable
- personal devotion to duty
- devoted to Army ethics
- genuine concern for others
- high concern for others
- caring concern for others
- impeccable, winning spirit
- promotes loyalty
- unselfish and trusting
- strength of character
- has high ideals, morals, and ethics
- unshakable character
- faces fear head-on
- ethical, honest personality
- strong ethical principles
- great sense of loyalty
- continually exhibits loyalty
- keeps composure under fear and stress
- strong advocate of Army ethics
- not easily excited under pressure
- not easily excited under stress
- possesses highest personal loyalty
- unbearable character and loyalty
- high morality and ethical principles
- even, steady temperament
- calm and composed during combat
- has high personal values
- unafraid of combat challenge
- positive unit spirit
- strong loyalty and sense of pride
- completely without bias or prejudice
- actively promotes soldiers' rights
- encourages professional pride
- encourages subordinates loyalty
- instills loyalty and pride

- contributes 110 percent to team effort
- promotes harmony and team work
- leader and member of the team
- encourages excellence among soldiers
- exhibits high trust level in soldiers
- dedicated to unit mission
- mission first, self last
- treats others with respect
- willing to share knowledge with soldiers
- demands maximum effort in team efforts
- stable and calm during "Operation Desert Storm"
- establishes team spirit throughout
- highly dedicated to unit goals
- devotion to duty and self-sacrifice
- a true team leader
- displays selfless devotion to duty
- unsurpassed devotion to duty
- unsurpassed devotion to soldiers' care
- self-sacrificing, a true team player
- has longstanding record of loyalty
- will not retreat in the face of fear

NEGATIVE VALUES STATEMENTS

- keeps growing professionally
- causes disorder and unrest
- disagreeable personality
- misrepresents the facts
- uncaring in matters of sympathy
- disruptive influence
- verbally abuses soldiers
- wears down subordinates spirit
- disobedient and belligerent
- incompatible, disagreeing personality
- incompetent without direct, consistent supervision
- unmanageable off-duty activities
- inclined to stray from the truth
- mood change without warning
- low level of self-confidence

- lacking in social graces and courtesy
- gives in to pressure during crisis situations
- cannot make proper decisions under pressure
- prejudicial to good order and discipline
- substandard duty performance
- unwillingness to conform to Army ethics
- overly apologetic and humble
- opposes all views other than own
- arbitrarily enforces standards
- lack of self-confidence
- requires close supervision
- fabricates the truth
- uncaring team leader
- cannot deal with stress and fear
- unwilling to lead during "Operation Desert Storm"
- shies away from soldiers' problems
- no pride in self or duty performance
- places self ahead of duty, unit and Army
- uncommitted to unit goals
- lackadaisical attitude
- short-tempered with soldiers
- untruthful to soldiers and leaders
- uninterested in Army ethics
- becomes insubordinate when talking to leaders

BULLET COMMENTS FOR COMPETENCE

POSITIVE COMPETENCE STATEMENTS

- demonstrates competent performance
- projects a special competence
- displayed competency during "Operation Desert Storm"
- demonstrates a high level of expertise
- demonstrates strong personal effectiveness
- demonstrates strong interpersonal competence
- excels in the effective application of skills
- displays a high level of technical competence
- blends technical skills with technical competence
- combines technical competence with loyalty

- demonstrates highly sophisticated skills
- processes specialized teaching skills
- processes specialized writing skills
- processes specialized speaking skills
- processes specialized listening skills
- processes specialized learning skills
- highly skilled in all phases of job
- highly skilled in all phases of training
- highly skilled in all phases of counseling
- displays excellent attention to technical skills
- is uniquely qualified to teach
- effectively capitalizes on strengths
- constantly sharpens and updates skills
- demonstrates professional expertise
- can be relied on to make sound decisions
- is willing to make unpopular decisions
- assembles facts before taking actions
- weighs alternative decisions before taking actions
- carefully evaluates alternative risks
- practices sound risk taking
- foresees the consequences of decisions
- makes sound decision under pressure
- avoids hasty decisions
- excels in profit-directed decisions
- concentrates on developing solutions
- excels in seeking solution
- makes decisions with confidence
- strives to improve decisiveness
- encourages decision with confidence
- demonstrates consistently distinguished performance
- ability to achieve desired results
- achieves bottom line results
- attains results through positive actions
- produces a tangible, positive impact
- exceeds performance expectations
- performance exceeds job requirements
- provides a competitive edge
- turns risk situations into opportunities
- possesses all traits associated with excellence

- uses all available resources
- generates enthusiasm
- is extremely resourceful
- works diligently until job completion
- displays concentrated effort
- displays trust and confidence
- faces conflicts with confidence
- completes tasks with confidence
- extremely self-confident
- maintains a high degree of involvement
- develops positive expectations
- develops realistic training
- sets high standards of personal performances
- writes in a positive tone
- writes reports that achieve maximum impact
- excels in making appropriate judgements
- can be entrusted to use good judgement
- effectively solves problems
- translates problems into workable solutions
- ability to solve problems
- diagnoses situations or conditions
- considers alternative courses of action
- exercises judgement on behavior of others
- shows eagerness and capacity to learn
- ability to learn rapidly
- benefits from all learning situations
- consistently strives to improve performance
- excels in self-supervision
- excels in self-improvement
- thirst for knowledge
- sound judgement
- quick to learn
- interesting convincing speaker
- advanced knowledge
- extensive knowledge
- firm, caring attitude
- exercises sound judgement
- capable of independent decision
- unlimited learning capacity

- relentless drive and ambition
- persuasive talker
- matchless desire
- self-motivated
- results oriented
- articulate speaker
- thinks and plans ahead
- eager and capable
- creative writing ability
- commands large vocabulary
- sound in thought, good in judgement
- skilled, eloquent speaker and writer
- unmatched appetite for learning
- strong ability to learn
- polished, persuasive speaker
- inspires self-improvement in soldiers
- always produces quality results
- learning growth potential
- unending drive to win
- unending drive and urge for success
- unquenchable thirst for knowledge
- continually seeks personal growth and development
- capable of independent thought and action
- has "follow me" confidence
- made sound judgements during "combat training"
- winning attitude during "Operation Desert Storm"
- set standards during "Operation Desert Storm"]

NEGATIVE COMPETENCE STATEMENTS

- poor planner, great hindsight
- lacking in knowledge
- lax in performance and behavior
- lacks in desire
- low self-esteem
- bad judgement
- serious judgement error
- shuns duty
- unpredictable work habits

- mild learning disability
- persistently poor performance
- does no more than required
- lack of pride in work
- lack of initiative
- negative attitude
- lack of pride in work
- exercises bad judgement
- fails to monitor subordinates
- lacks self-discipline
- sloppy workmanship
- frequent bad judgement
- gives misguided direction
- obvious lack of self-motivation
- gets less than required results
- fails to achieve consistency
- lack of confidence in abilities
- lacks knowledge and ability
- lacks consistency in performance
- deficient in skill and knowledge
- incapable of sustained satisfactory performance
- uncertain in making decisions
- doubtful in making decisions
- lacks good judgement and common sense
- unwilling to listen to soldiers
- avoids duty during prime time training
- written products are not clear and coherent
- blunt and rude in speech and manner
- unable to choose correct course of action
- shows no desire for improvement
- exhibits only short periods of success
- unable to distinguish right from wrong
- decisions are open to question
- fails to achieve minimum acceptable performance
- makes careless, avoidable mistakes
- opposes improvement efforts of other
- cannot articulate in speech
- incoherent in writing
- dodges work and responsibility

- not job-aggressive
- relies too heavily on others
- below average performer
- sometimes premature in judgement
- best efforts prove fruitless
- best efforts prove unsuccessful
- speaks and acts on impulse
- plagued by lack of self-confidence
- inattentive to details
- slow and deliberate work pace
- performance is all valleys and no peaks
- performs well below peer group
- less than marginal performer
- seriously lacking in initiative
- lacks counseling skills
- total lack of enthusiasm
- reluctance to conform to standards
- unsophisticated reasoning and judgement
- slow to learn and develop
- failed to achieve consistency in training
- lacks the capacity to lead
- bad leader, good follower
- loses composure when under stress
- unsuitable for NCO corp
- sometimes lax in supervision
- not serious about training soldiers
- has no idea how to train soldiers
- unable to train soldiers
- CTT scores lower than his soldiers
- has difficulty with training soldiers
- planned badly for SDT
- makes hasty decisions

BULLET COMMENTS FOR PHYSICAL FITNESS AND MILITARY BEARING

POSITIVE FITNESS STATEMENTS

- grasps the most difficult concepts

- exceptionally keen and alert
- reasonable, smart and keen
- alert, quick and responsive
- sustains a high level of concentration
- independent thinker
- thinks before taking action
- thinks fast on feet
- uses common sense
- uses sound fact-finding approaches
- demonstrates innovative insight
- uses intelligent reasoning
- displays strong memory skills
- displays strong mental flexibility
- possesses strong memory skills
- thinks futuristically
- effectively manages stress
- performs well under pressure
- gets things done calmly
- recognizes the importance of appearance
- presents an attractive appearance
- takes pride in personal appearance
- grooming is neat, attractive and appropriate
- master at personal hygiene and dress
- conforms to proper standards of dress
- dresses appropriately for all occasions
- projects a positive image
- perceptive and alert
- composed and calm
- great mental grasp
- a quick thinker
- always enthusiastic
- emotionally stable
- mentally alert
- mentally sharp
- bold, forward thinker
- calm and affable manner
- powerful, influential figure
- alert, energetic personality
- sound of mind and judgement

- pleasing personality
- analytical mind
- boundless energy
- intellectual courage
- strong will of mind
- sharp, mental approach
- quick, penetrating mind
- a pillar of strength
- 10K running team member
- post basketball team member
- mentally and physically able
- scored 300 on APFT
- another 300 APFT score
- only E-7 on post to score 300 on APFT
- awarded the Army Physical Fitness Patch
- outscored all peers in BN on APFT
- outscored his platoon during APFT
- coach for unit football team
- in charge of overweight soldiers in unit
- teaches stress management to peers
- unit stress management NCO
- organized unit "quit smoking" campaign
- organized training for overweight soldiers
- meticulously well-groomed
- uses common sense to tackle problems
- exceptional personal drive and energy
- appearance, without equal
- positive mental attitude and outlook
- mentally quick and active
- strong, positive drive
- a pillar of moral strength and courage
- dignified in presence and appearance
- physically ready, mentally alert
- mannerly, courteous, and polite
- displayed physical vigor during BN "Organization Day"
- strong will of mind and commanding presence
- impressive posture and appearance
- meets Army standards by APFT
- always active, on or off duty

- great mental aptitude
- a perfect example of physical fitness

NEGATIVE FITNESS STATEMENTS

- not physically fit by Army APFT standards
- lacks physical vigor
- weak, inadequate leader
- not well-organized mentally
- emotionally immature
- failed last APFT
- careless appearance
- lacks military bearing
- failed to improve low APFT scores
- lack of desire for physical training
- loses emotional control
- lacks proper mental discipline
- does no more than required during APFT
- sets bad example for physical fitness
- lags behind others during PT run
- failed to live up to expectations of a leader
- lacks mental depth and soundness
- not worthy of present rank
- sometimes lax in physical training
- sometimes lax in appearance
- military bearing is below standards
- substandard military bearing
- has bad habit of skipping morning PT

BULLET COMMENTS FOR LEADERSHIP

POSITIVE LEADERSHIP STATEMENT

- performs with a high degree of accuracy
- performs with consistent accuracy
- strives for perfection
- excels in achieving perfection
- believes in bottom-up leadership
- meets precise standards

- communicates high expectations
- communicates clearly and concisely
- demonstrates sound negotiating skills
- creates opportunities
- initiates fresh ideas
- is very dependable and conscientious
- extremely reliable and supportive
- successfully builds soldiers
- trains soldiers to become leaders
- facilitates learning
- builds on strengths
- identifies soldiers needs
- plans for effective career development
- seeks personal growth and development
- understands personal strengths and weaknesses
- encourages broad development of soldiers
- assists soldiers with skills, knowledge and attitudes
- sets reachable targets
- sets reachable goals
- sets worthy goals
- is a goal seeker
- effectively establishes group goals
- achieves and surpasses goals
- shows concern for soldiers
- believes in soldiers care
- shows concern for soldiers development
- concern about professional development
- created study plan for NCO's SDT
- takes charge without being told
- always ready to take over the next leadership position
- radiates confidence
- inspires confidence and respect
- projects self-confidence
- demonstrates natural leadership ability
- demonstrates strong, dynamic leadership
- ability to stimulate others
- commands the respect of others
- knows how to get soldiers' attention
- maintains a mature attitude

- speaks with a positive tone
- speaks with a pleasant tempo
- excellent persuasive abilities
- develops sound contingency plans
- develops positive expectations
- expects and demands the best
- committed to excellence
- strives for perfection
- performs at peak performance
- performs at a high energy level
- exceeds normal output standards
- understandS human behavior
- easily gains acceptance of others
- best impression in every situation
- conveys a positive personal image
- gets along well with others
- builds a close rapport
- promotes harmony among soldiers
- interacts effectively with soldiers and peers
- conveys positive influences
- excels in promoting team efforts
- establishes realistic work demands
- encourages soldiers to win
- effectively balances work flow
- builds cooperation
- promotes positive involvement
- stimulates individual participation
- gives helpful guidance to new soldiers
- shows a sincere interest in soldiers
- excels in effective counseling of soldiers
- effectively uses counseling techniques and skills
- gives sound, practical advice
- properly asserts authority
- gains soldiers confidence
- shows empathy
- shows genuine respect
- sensitive to the feelings of others
- supervises firmly and fairly
- is fair and firm with soldiers

- turns complaints into opportunities
- quickly settles disciplinary problems
- corrects without criticizing
- takes appropriate remedial action
- tactful in conflict situations
- negotiates with tact
- fighting enthusiasm
- always ahead of the action
- good samaritan
- dedicated to helping others
- dedicated to soldiers' care
- result-oriented individual
- unlimited capacity for solving problems
- highly motivated achiever
- always a willing volunteer
- motivates and leads others
- a real motivator
- composed leader
- good organizer
- firm, resolute leader
- bolsters spirits
- inspires performance
- good-natured team leader
- positive motivator
- firm, yet fair leader
- respected by peers and other leaders
- uncommon leadership
- inspires greatness
- promotes sound leadership
- accomplished counselor
- skillful, direct leadership
- astute, experienced leader
- unparalleled leadership
- motivates and leads others
- well-rounded leadership skills
- stirs up enthusiasm
- charismatic leader
- no-nonsense leader
- inquisitive leader

- impressive leadership
- selfless leader
- team leader
- concerned, caring leader
- has personal leadership magic
- sensitive to the needs of others
- frank, direct leader
- an accomplished counselor
- vigorous leadership style
- promotes *esprit de corps*
- impressive leadership record
- tactful leader and motivator
- engenders self-development
- exercises sound leadership principles
- exceptional leadership
- influence soldiers to accomplish the mission
- provides purpose, direction and motivation
- correctly assesses soldiers' competence
- correctly assesses soldiers' motivation
- correctly assesses soldiers' commitment
- takes proper leadership action
- creates climate that encourages soldier participation
- develops mutual trust, respect, confidence in soldiers
- is aware of his strengths, weaknesses and limitations
- ensures soldiers are treated with dignity and respect
- considers the available resources
- considers subordinates level of competence
- skillful in identifying/thinking through the situation
- takes quick corrective action
- seeks self-improvement
- technically and tactically proficient
- seeks and takes responsibility
- makes sound and timely decisions
- leader in soldiers' care
- keeps soldiers informed
- develops a sense of responsibility in soldiers
- ensures tasks are understood, supervised, accomplished
- develops highly effective team
- knows limitations and capabilities of soldiers

- accomplishes all assigned tasks to the fullest
- willingness to accept full responsibility
- honest and upright
- sincere, honest, candid and avoids deceptive behavior
- explains the "why" of the mission
- teaches soldiers to think creatively
- solves problems while under stress
- able to plan, maintain standards and set goals
- supervises, evaluates, teaches, coaches, and counsels
- rewards soldiers for exceeding standards
- serves as ethical standards bearer
- develops cohesive soldier teams
- corrects performance not meeting standards
- leader with strong and honorable
- a role model for all to follow
- committed to Army ethics
- chooses the right course of action
- frank, open, honest and sincere
- competent and confident leader
- establishes broad categories of skills and knowledge
- able to understand and think through a problem
- ability to say the correct thing at the right moment
- does not interrupt when others are speaking
- ensures soldiers are professionally and personally developed
- ensures the tasks are accomplished
- influences the competence and confidence of soldiers
- helps soldiers develop professionally and personally
- understands how soldiers learn
- motivates soldiers to learn
- an expert in teaching and counseling
- counsels soldiers frequently
- counsels soldiers on strengths and weaknesses
- good at identifying and correcting problems
- understands human nature
- has great listening skills
- leads soldiers to making their own decisions
- knows when to be flexible and unyielding
- creates strong bond with soldiers
- builds soldiers' spirit, endurance, skills and confidence

- works hard at making soldiers team members
- knows war fighting doctrine
- operates and maintains all assigned equipment
- not afraid to ask seniors, peers, or soldiers for help
- understands doctrine and tactics of potential enemies
- gathers facts and makes assumptions
- develops possible solutions
- analyzes and compares possible solutions
- recognizes and defines the problem
- selects the best solutions
- able to make timely decisions
- identifies the best course of action
- involves soldiers in making decisions
- understands and uses backward planning
- develops a schedule to accomplish task
- recognizes and appreciates soldiers abilities
- uses computers, analytical techniques, and other technological means
- good example setter
- corrects performance not meeting standards
- makes soldiering meaningful
- makes decisions soldiers accept
- plans and communicates effectively
- an example of individual values
- learns leadership skills from other leaders
- teaches soldiers to depend on each other
- develops full potential of soldiers
- constantly stresses teaching, coaching and caring
- constantly stresses bonding, learning, and teaching
- creates trust and strong bonds with and among soldiers
- strong desire to excel
- inspires zeal and obedience
- displays concern for soldiers
- experienced, knowledgeable leader
- ability to inspire others
- stimulating leader
- a "leader by example" who obtains superior results
- a dynamic, motivating leader who obtains superior results
- strict, firm disciplinarian

- animating leader, arouses enthusiasm
- supervises and directs soldiers
- instills pride and dignity in others
- strong, decisive leader
- dynamic leader and vigorous worker
- dedicated to unit's mission and goals
- hard-line leader with unyielding character
- provides assistance to those in need
- potent, productive leader
- vigorous leadership, superb management
- capitalized on soldiers' strengths
- epitome of tactful leadership
- stimulates subordinates professional growth
- intense, compassionate leader
- invigorating, successful leader
- brings out best in soldiers
- gives support to chain of command
- informed leader who cares about solders
- concerned, caring leader
- compassionate, caring leader
- considerate for the feelings of others
- shares time and knowledge with soldiers
- offers positive advice to soldiers
- a demanding leader who gets impressive results
- demonstrates superb leadership
- work-aggressive leadership
- radiant, confident leader
- highly respected leader and organizer
- superb leadership
- master of superb leadership
- proven leader of unbounded ability
- imaginative leadership techniques
- poised and mature leader
- assertive and considerate leader
- proponent of strong, solid leadership
- persuasive and tactful leader
- leads with intensity, force, and energy
- develops soldiers at a rapid pace
- encourages off-duty professional growth

- enforces rules and regulations
- personifies leadership by example
- gives a pat on the back when earned
- knows key to quality performance
- recognizes and rewards top performances
- ignites enthusiasm throughout the ranks
- effective leadership qualities
- unfailing devotion to duty and country
- gives praise when most deserved or needed
- equality and equitable treatment of soldiers
- a real motivator and dedicated leader
- encourages each soldier to set high goals
- leads soldiers to desired level of performance
- demands the best from soldiers
- displays tactful leadership daily
- combat leader
- gives subordinates demanding responsibility
- leader of uncommon perceptiveness
- stands above peers
- high achiever
- unblemished record
- walking FM 22-100
- a role model for all NCO's accustomed to success
- a top professional
- committed to excellence
- seeks challenging assignments
- sets professional example
- overcomes all obstacles
- superior to peers
- head and shoulders above peers
- reliable and dependable
- skilled in art of leadership
- highly specialized in leadership
- made marked improvement during past year
- a champion in the field of leadership

NEGATIVE LEADERSHIP STATEMENTS

- lacking in knowledge

- dispassionate leader
- weak, inadequate leader
- impersonal leader
- inferior performer
- inconsiderate and uncaring
- dull, uninspiring leader
- not a potent, effective leader
- weak, ineffective leader
- lack of confidence
- leadership vacuum
- insensitive leadership
- unimpressive leader
- inflexible leader
- concerned with own self-interest
- lags behind others
- apathetic leader
- suppresses subordinates growth
- inflexible, unimaginative leadership
- deficient in skills and knowledge
- failed to live up to expectations
- irresponsible leadership
- impatient with subordinates
- bad leader and worse follower
- insensitive and calloused leader
- not a strong leader
- too permissive and lenient
- careless of the feelings of others
- suitable for routine, ordinary jobs
- unforgiving leader, rubs in mistakes not out
- leadership style leaves uneasy feeling
- non-forgiving leader, doesn't forget mistakes
- leadership ability is questionable
- total lack of enthusiasm
- loses composure when under pressure
- stirs up resentment
- over supervises soldiers
- breeds resentment and low morale
- lack of coordination
- disorganized leader

- sets bad examples for soldiers
- a leader that should not be
- can't lead more than one soldier
- wants credit but can't lead
- a leader with no know-how
- last NCO to leave in charge
- can't be trusted to accomplish the mission
- soldiers tell him what to do

BULLET COMMENTS FOR TRAINING

POSITIVE TRAINING STATEMENTS

- a great raw ability and talent
- stimulating intelligence
- powerful, influential figure
- advanced knowledge
- unlimited teaching capacity
- sparks excitement
- personal devotion to training
- winning spirit
- train to changing situations
- dedicated to helping soldiers learn
- teach from combat experience
- realistic trainer
- superb trainer and leader
- unmatched capacity for teaching
- strong desire and ability to teach and train
- persuasive trainer and supervisor
- trains soldiers to be trainers
- teaching and training ability is unlimited
- always coming up with something better
- stands above contemporaries in training soldiers
- always willing to teach and train
- most professional talent is training
- a leader in training soldiers
- a take charge attitude with great training ability
- shares knowledge with others
- possesses an impressive breadth of experience

- continually strives for professional development
- promotes professional growth
- plans ahead for training success
- persuasive, convincing trainer
- strong will and know how
- relentless drive and dedication
- excellent teaching ability
- uses own initiative to train soldiers
- possesses abundance of enthusiasm and drive
- stands up for principles and beliefs
- devotion to duty always
- teaches soldiers how to survive during combat
- real morale booster and trainer
- trains soldiers to win and survive
- arouses and excites when training
- accomplished trainer
- skillful, direct trainer and leader
- inspires and encourages during training
- well-rounded training skills
- stirs up enthusiasm
- no-nonsense trainer
- impressive trainer
- trains so all can learn
- fully taxes soldiers when training
- inspires the imagination
- positive influence, great motivator
- knows success in team training
- vigorous training style
- impressive training record
- enforces soldiers' participation
- has ability to inspire others to train
- experienced, knowledgeable trainer
- dedicated to betterment of subordinates
- contributes maximum effort and energy
- invigorating trainer and leader
- excites and arouses others to action
- training style elicits harmony and team work
- dedicated to mission training
- brings out the best in subordinates

- effectively trains and directs inexperienced personnel
- skillful in training others to desired goals
- training merits special praise and gratitude
- performance oriented trainer
- trains to sustain proficiency
- trains to maintain
- trains to achieve combat-level standards
- trains soldiers to fight
- effectively uses realistic conditions when training
- includes simulations, simulators, and devices in training
- trains to challenge
- his training excites and motivates soldiers
- sustains skills to high standards
- operates in a "band of excellence"
- uses all available time for training
- responsible for training and performance of soldiers
- bases training on wartime mission requirements
- provides the required resources for training
- helps develop mission training programs
- awarded for best mobilization plan
- responsible for the best maneuver team
- develops supporting task list for each METL
- maintains consistent training awareness
- concerned with future proficiency
- effectively uses all available resources
- allocates maximum training time
- provides training recommendations for unit training
- helps improve soldiers warfighting skills
- provides specific guidance to trainers
- prepares detailed training schedules
- creates a bottom-up flow of information regarding training
- specifies tasks to be trained
- provides concurrent training topics for training
- specifies who conducts and evaluates training
- provides training feedback to commander
- personally observes and evaluates all training
- demands training feedback from subordinate leaders
- executes and evaluates training
- coaches trainers on how to train

- provides time to rehearse training
- competes to achieve the prescribed standards
- trains evaluators as facilitators
- creates well-trained and highly motivated soldiers
- is honest about wartime situations
- guides team towards mission accomplishment
- listens and responds fairly to criticisms
- guides soldiers into accepting team goals
- sets realistic goals
- periodically checks on team progress and training
- concerned about soldiers' safety and survival in combat
- a catalyst of teamwork and high morale
- willing to share knowledge and talent
- promotes maximum effort in team training
- first to learn, last to forget
- a proven trainer of unbounded ability
- bold and imaginative training techniques
- superior knowledge of training and teaching
- always willing to train others
- persuasive and tactful trainer
- trains with intensity, force, and energy
- realistic, motivated trainer
- spent extensive hours developing realistic training plans
- never to busy to train soldiers
- unique ability to coordinate group efforts
- popular among peers and soldiers
- mold team into cohesive, productive unit
- establishes and enforces clear-cut goals
- demands positive results
- successful conclusion never in doubt
- trains team members to take charge
- trains team members to train each other
- true team player
- plans carefully and wisely
- trainer of trainers
- primary trainer for unit
- trains to win and succeed
- trains and tests results

NEGATIVE TRAINING STATEMENTS

- dull, uninspiring trainer
- unable to master training skills
- loses control when training
- lack of confidence when training
- ignores reality
- pressures soldiers when training
- throws weight around when training
- wasted training opportunities
- deviates from trainer standards
- unimpressive trainer
- dull, trite personality
- undermines morale
- irritates and annoys others
- discourages team unity
- his soldiers are better trainers
- becomes easily frustrated
- lack of initiative
- fails to monitor training
- lack of personal conviction
- blames others for own short-comings
- impatient with soldiers
- fails to achieve minimum performance
- fails to train soldiers during prime time
- lacking in ability to train
- bad habits include not training soldiers
- shies away from training
- depends on others to train his soldiers
- takes short cuts in training
- his training leads to failure

BULLET COMMENTS FOR RESPONSIBILITY AND ACCOUNTABILITY

POSITIVE RESPONSIBILITY STATEMENTS

- knows whereabouts of soldiers at all times
- accepts responsibility of own decisions

- accepts responsibility of soldiers' actions
- accepts ultimate responsibility
- assumes responsibility for mistakes
- never runs from responsibility
- effective in assigning responsibility
- delegates responsibility effectively
- devotes attention to responsibility
- meets responsibility
- fulfills all commitments
- meets all schedules and deadlines
- can be counted on to achieve positive results
- consistently punctual
- effectively follows up assignments
- responsible leadership
- takes responsibility for right or wrong
- takes responsibility for good or bad
- understands and takes responsibility
- exceptional responsible leader
- responsible for team morale, strength and courage
- responsible and enthusiastic leader
- takes responsibility for soldiers' safety
- enforces responsibility and accountability
- seeks responsibility for self and soldiers
- responsible for team success
- prompt and responsive
- responsibilities discharged superbly without fail
- actively seeks additional responsibility
- thrives on important responsibility
- appetite for increased responsibility
- carries out responsibility in a professional manner
- accepts added responsibility without wavering
- aggressive in assumption of additional responsibility
- gives whole-hearted support to accountability
- assumed expanded responsibilities
- directly responsible for four million dollars worth of equipment
- responsible for upgrading soldiers CTT scores

NEGATIVE RESPONSIBILITY STATEMENTS

- not aware of soldiers whereabouts
- lost valuable equipment during FTX
- shirks responsibility
- avoids work and responsibility
- didn't know equipment was missing
- will not stand up for soldiers
- does not seek responsibility
- can't count on to lead or train
- not responsible
- not accountable
- can't be counted on to lead
- must do one thing at a time
- can't count on to care for soldiers
- wasn't aware of missing equipment
- wasn't aware his soldier was AWOL
- late for duty more than his soldiers
- forgot soldier was waiting for his inspection

BULLET COMMENTS FOR OVERALL POTENTIAL FOR PROMOTION

POSITIVE PROMOTION STATEMENTS

- promote now
- should have been promoted yesterday
- don't hold back promotion
- promote before peers
- on the road to CSM
- sergeant with first sergeant potential
- promotion will benefit the Army
- has CSM potential
- promote yesterday
- promote ahead of peers
- highest recommendation for promotion
- whatever it takes, promote
- promotion will justify performance
- uniquely qualified for CSM

- on the road to the top
- been looked over too long, promote
- none better for next higher grade
- consistently strives to improve performance
- excels in self-improvement
- should serve as first sergeant
- promotion overdue
- look over TIS and promote
- plans for effective career development
- makes valuation suggestions for improvements
- sets ambitious growth goals
- displays an eagerness to improve
- welcomes opportunities for improvement
- high degree of professional excellence
- a self-starter
- finds new and better ways to perform duties
- does things without being told
- requires no supervision
- excels in solving critical problems
- well-informed on job and social issues
- always ready to take charge
- extremely dedicated, and success-oriented
- displays intense involvement
- work in a higher grade positions
- works in an E-__ position
- present position calls for an E-__
- gives maximum effort
- strong achievement drive
- maintains self-motivation
- consistent production of high quality of work
- highest standards of excellence
- quality of work reflects high professional standards
- performs at peak efficiency
- reads at a CSM level
- writes at a CSM level
- completed speed reading course
- promotes team efforts
- seeks the hard assignments
- expects and demands superior performance

- gains maximum productivity from soldiers
- develops a spirit of teamwork
- outperforms peers three to one
- seeks hardship assignments
- optimistic outlook and attitude
- good academic aptitude
- reaches success by self-motivation
- fully capable of the next higher grade
- leadership ability is unlimited
- special talent for areas of next grade
- supervises firmly and fairly
- takes charge, a true leader and motivator
- dynamic leader with combat experience
- contributes 110 percent
- leadership merits special praise and gratitude
- leadership merits promotion to the next highest grade
- rewards superior performance with promotion
- not promoting him is giving away stripes
- knows key to quality performance is leadership
- acts decisively under pressure
- provides timely advice and guidance
- sets standards by which excellence is measured
- dedicated to mission purpose
- exceptionally well-organized
- succeeds despite adversity
- self-sacrificing, duty first
- a model for all to emulate
- seeks opportunities to grow professionally
- gives serious and determined effort
- almost infinite growth potential
- well-rounded and professionally knowledgeable
- possesses all attributes required to excel to the top
- supports and enforces all rules and orders
- promptly executes all orders
- sets and achieves high goals
- realizes success requires sacrifice and dedication
- always involved in something constructive
- especially strong in the execution of demanding tasks
- did a masterful job during—

- a foremost authority on combat training
- without equal in ability to lead
- a top specialist in the field of—
- stands out in the field of—
- surpasses peers in sheer ability to—
- has the knowledge and competence to—
- widely respected for ability to—
- rejuvenated and put new life into training
- exhibits all essential features of a CSM
- masterful ability to lead junior NCO's
- a remarkable skilled NCO
- impressive accomplishments include training
- rare, extraordinary ability to train junior leaders
- skilled in art of counseling
- thriving force behind success
- thoroughly understands soldiering
- maintains sharp edge in leading others
- an absolute master at training leaders
- uniquely skilled to get things done
- has veracious appetite to be the best
- no end to potential for promotion

NEGATIVE PROMOTION STATEMENTS

- professionally stagnated
- do not promote
- candidate for QMP
- promote after peers
- make last on promotion list
- needs to earn present rank
- promotion would be a gift
- not ready for next stripe
- working towards promotion
- hold promotion until next year
- should be reduced, not promoted
- put promotion on hold
- totally unconcerned about more responsibility
- lack of desire
- lags behind others in his MOS

- should move back, not forward
- no future potential
- lax in carrying out duties
- a burden to leadership
- unable to overcome minor problems
- devoid of hope for improvement
- expects success using guesswork and conjecture
- may come around in due time
- does not have the ability to lead
- not qualified to be an E-___, let alone an E-___
- incompetent in area of _______ planned badly for career progression
- has not kept pace with peers
- please don't promote
- promotion would be a waste
- if no way out, promote
- save promotion for new peers

BULLET COMMENTS FOR OVERALL PERFORMANCE

POSITIVE OVERALL STATEMENTS

- achieves bottom-line results
- attains results through positive actions
- produces tangible, positive actions
- performance exceeds job requirement
- provides a competitive edge
- possesses all traits associated with excellence
- displays many areas of strength
- is extremely resourceful
- works diligently
- displays strong perseverance
- competes with confidence
- high degree of involvement
- high standards of personal performance
- possesses personal magnetism
- quality of work is consistently high
- excels in getting work done by others

- expects and demands superior performance
- effectively uses counseling techniques and skills
- effectively deals with misunderstandings
- accomplishes results without creating friction
- concentrates on area in need of help
- demonstrates competence in many areas
- ability to perform a wide range of assignments
- effectively handles special assignments
- copes with accelerating changes
- totally committed to excellence
- intense dedication to duty
- stands above contemporaries
- stands above most seniors
- positive, fruitful future
- highest standards of excellence
- seeks challenging hardship assignments
- overcomes all obstacles
- ace technician and leader
- superior to others
- impressive accomplishments
- quick to take positive actions
- reaches full potential quickly
- total, complete professional
- uncommon excellence
- expects total effort
- highly competent and dedicated
- demands quality performance from all
- true meaning of "pride and professionalism"
- thrives on responsibility
- exceeds highest professional standards
- a front-runner in every category
- considers no job too difficult
- trains for the unexpected
- a leader of his peers
- performs beyond professional abilities
- at the pinnacle of professional excellence
- a zealot in performing any assignment
- takes pride in doing best job possible
- symbolizes the top quality professional

- sets standards by which excellence is measured
- executes tasks expediently and correctly
- enhances and improves morale
- impressive record of accomplishments
- sets and achieves high personal standards
- performance exceeds job requirements
- highly skilled in all phases of job
- always productively employed
- unfailing performance to duty
- achieves quality results
- puts forth unrelenting effort
- workload is correctly balanced and prioritized
- exceptionally high level of performance
- sets the pattern and example for peers
- deeply devoted to chosen profession
- frequently seeks out expert opinion
- first-rate professional
- rates first against any competition
- at the forefront of peer group

NEGATIVE PERFORMANCE STATEMENTS

- downward turn in performance and attitude
- substandard performance across the board
- overall performance is below standard
- requires close and constant supervision
- does not promote good morale
- average leadership skills
- failed SDT
- performance declined during rating period
- goes out of way to irritate others
- a complete disappointment
- ignores direction and guidance of superiors
- no desire for improvement
- disrespectful attitude and behavior
- work marked by others, complete failure
- difficult for others to work with
- suitable for routine, ordinary jobs
- tries hard but achieves little

- leadership school was no help
- overly concerned with self-image
- far below average performer
- poor leader and follower
- leadership ability is questionable
- inattentive to details
- plagued by indecisiveness
- duty performance totally unsatisfactory
- does nothing right
- unreliable performance despite counseling
- performance far below peers
- incompetent in all areas of leadership
- careless and negligent in duty performance
- attitude and performance not in tune with pay grade
- has bad habit of being late
- has no idea how to lead
- has difficulty with team training
- planned badly for AGI
- ignores advice of superiors
- causes extra work for others
- continuing discipline problem
- indecisive under pressure
- lags behind others
- don't care attitude

There you have it, a list of words to use to form your bullet comments as well as a list of bullet comments that you can add to or take from to build the comments you desire. You can even use the words and comments to write:

- Letters of appreciation;
- letters of recommendation (promotions);
- letters of commendation; and
- narrative for awards.

Remember, there are some weak leaders that feel they have control over you because they are your rater. Don't let that intimidate you. All you have to do is make sure you do your job the best you can and make sure you are counseled and understand what you were counseled about.

Should you receive a bad NCO-ER before you were able to do what is outlined in this book, read "Appealing the NCO-ER" in the next chapter.

In Chapter Four, you can also read about the types of NCO-ER's and counseling techniques. Here, we also will examine the Self Development Test.

Chapter Four

HELPFUL INFORMATION

TYPES OF NCO-ER's

There are five types of NCO-ER's that you may be associated with during your stay in the Army. They are:

- First report;
- Annual report;
- Change of Rater report;
- Complete the Record report; and
- Relief for Cause report.

Three of the reports listed above are mandatory because they are required to be used for all Sergeant, E-5's, to Command Sergeant Majors. These reports are the:

- Annual report;
- Change of Rater report; and

- Relief for Cause report.

The Relief for Cause report is a mandatory report but only if the NCO gets relieved from duty.

The Complete the Record report is one very few NCO's get, however, I will tell you more about it later.

The First Report, which is not a report in itself, is the first report the NCO will receive. It will be determined by the date of the event requiring a report (eg., change of rater, annual). The beginning month will be the month of the effective date of promotion to sergeant, reversion to NCO status after serving as a commissioned or warrant office for 12 months or more, re-entry on active duty after a break in service of 12 months or more, or the date of the ABCMR memorandum which approves reinstatement of the NCO rank. You will only receive one First Report unless you ETS and come back into the Army or become an officer or a warrant officer and, for some reason, become a NCO again. Your First Report will most likely be your first Annual or Change of Rater report. Your Annual Report will be submitted 12 months after the most recent of the following events:

- ending month of last report;
- effective date of promotion to sergeant;
- reversion to NCO status after serving as an officer or warrant officer for 12 months or more;
- re-enter on active duty in a rank of sergeant or above after a break in enlisted service 12 months or more.

You must work for your rater for a minimum of three months before you can get an Annual Report. If not, the reporting period will be extended until the minimum rater qualifications are met.

The senior rater will complete both the rater's and senior rater's portions of the NCO-ER if the rater dies or if the reviewer, on the advice of medical authorities, believes the rater is unable to submit an accurate evaluation because he was:

- relieved from duty;

- reduced in rank;
- absent without leave;
- declared missing; or
- incapacitated.

The senior rater must be in the direct line of supervision of the rated NCO for a minimum period of two months if he has to do the rater's and senior rater's parts of the NCO-ER.

An Annual Report will not be signed prior to the first day of the month following the ending month of the report.

The Change of Rater report is submitted whenever the designated rater is changed as long as minimum rater qualifications are met. The Change of Rater report is also submitted when the:

- rater or rated NCO are reassigned;
- rater is attending a resident course of instruction or training scheduled for 90 calendar days or more at a service school;
- rater is attending a civilian academic or training institution on a full-time basis for a period of 90 calendar days or more;
- rater is performing duties not related to his primary function in his parent unit and is under a different immediate supervisor for three rated months or more;
- rater or rated NCO is released from active duty early of normal ETS, except for discharge and immediate reenlistment;
- rated NCO is reduced to SPC or below (reduction to another NCO grade does not require a report unless the actual rater changes); or
- rater dies, is relieved, AWOL, declared missing, or becomes incapacitated to such an extent that the reviewer believes the rater is unable to submit an accurate evaluation.

When both the rater and senior rater are unable to evaluate the NCO because of any combination of these factors, a report will not be submitted. The period will be shown as non-rated on the next report.

Code "Q" will be used to explain the non-rated periods. A NCO that is attached to another organization pending compassionate reassignment remains the responsibility of his parent unit and will not receive a NCO-ER from the attached unit. Also, a NCO that is on TDY or special duty and remains the responsibility of the rating officials in his parent unit will continue to be rated for that period, regardless of its length. An NCO that is on TDY or special duty, and is not the responsibility of the rating officials in his parent unit, will be rated by his TDY or special duty supervisors. A Change of Rater report will be submitted when requested by the rater or rated NCO upon approved retirement (ending month will be the month terminal leave begins). The Change of Rater report may not be signed before the date the change occurs but, in the event of a PCS, the report may be completed and signed up to 10 days prior to the date of departure in order to facilitate orderly outprocessing.

The Complete the Record report may be submitted on a NCO who is about to be considered by a DA centralized board for promotion, school, or CSM selection, providing the following conditions are met.

- The rated NCO must be in the zone of consideration (primary or secondary) for a centralized promotion board or in the zone of consideration for a school or CSM selection board.
- The rated NCO must have been in the current duty position under the same rater for at least six rated months as of the ending month established in the message announcing the zones of consideration.
- The rated NCO must have not received a previous NCO-ER from the same rater for the current duty position.The Complete the Record report is an optional report and will not be a basis for request for stand-by reconsideration. The report will not be signed prior to the first day of the month following the ending month.

The last NCO-ER I would like to tell you about is the Relief for Cause report. This is one report that you do not want to receive. Remember, should you be relieved from duty for not doing your job, you should have received counseling statements concerning the facts.

Relief for Cause is defined as the removal of a NCO from a rateable assignment based on a decision by a member of the NCO's chain of command or supervisory chain that the NCO's personal or professional characteristics, conduct, behavior, or performance of duty warrant removal in the best interest of the U.S. Army (AR 600-200). If Relief for Cause is contemplated on the basis of an information AR 15-6 investigation, the referral procedures contained in that regulation must be complied with before the act of initiating or directing the relief. This does not preclude a temporary suspension from assigned duties pending application of the procedural safeguards contained in AR 15-6.

The bullet comment, "rated NCO has been notified of the reason of relief" will be entered in Part IV(f) by the rater. If the relief is directed by an official other than the rater or senior rater, the official directing the relief will describe the reason for relief in an enclosure (not to exceed one page) to the report and remember the NCO must be provided a copy of the completed report including any authorized enclosures.

The minimum rating period of a Relief for Cause report is 30 days. The fundamental purpose for this restriction is to allow the rated NCO a sufficient period to react to performance counseling during the rating period. Authority to waive this 30 day minimum period in clear-cut cases of misconduct is hereby granted to a general officer in the chain of command or an officer having general court martial jurisdiction over the relieved NCO. The waiver approval will be in memorandum format and attached as an enclosure to the report. It will be prepared as outlined in AR 340-15 on 8 1/2 by 11 inch paper. When the rater is relieved, or when the rated NCO and the rater are concurrently relieved, the senior rater will complete both the rater's and senior rater's portions on each of the rater's subordinates. On DA Form 2166-7 "rater relieved" should be entered in Part V(e), the relieved NCO should not be identified in Part II(a), and Part 1, Item (i) will reflect one rated month, the date of relief will determine the "thru" date of the report. Relief for Cause reports may be signed at any time during the closing of the following month of the report.

That takes care of the information about the types of NCO-ER's.

Now, I want to share information with you about something I hope you will never have to do, appeal the NCO-ER.

APPEALING THE NCO-ER

A bad NCO-ER is one that has one or more Needs Improvement bullet comments. Should you get a report with a Needs Improvement comment, you should first talk with the person that gave it to you and try to get it changed. If that does not work, talk to the reviewer, but do this only if you feel you should have received a better NCO-ER.

If all else fails and you receive the bad report, or it looks like you will, the first thing you may want to do is make an objective analysis of the NCO-ER. Talk to the PSNCO or someone you feel is a true leader to determine whether an appeal is advisable.

The next step you may want to take is to request a commander's inquiry. The commander's inquiry will be conducted by a commander (major or above) in the chain of command above the designated rating official(s) involved in the allegations. The commander will confine the inquiry to:

- Matters relating to the clarity of the report;
- The facts contained in the report;
- The compliance of the report (IAW AR 623-205);
- The conduct of the rated NCO and rating officials.

The procedures for the inquiry may be as formal or informal as the commander deems appropriate, to include telephone and personal discussions. The commander may also appoint an officer, senior to the designated rating officials involved in the allegations, to make the inquiry. The primary purpose of the commanders's inquiry is to provide a greater degree of command involvement and preventing obvious injustices to the rated NCO and correcting errors before they become a matter of permanent record. A secondary purpose is to obtain command involvement in clarifying errors or injustices after the NCO-ER is accepted at USAEREC, a state adjutant general's office, or APPERCEN. However, the commander's inquiry is not intended to be a substitute for the appeals process, which is the primary means of addressing errors and injustices, after they have become a matter of permanent record. The commander's inquiry procedure will not be used to document differences of opinion between rating officials, or between the commander and rating officials about an NCO's performance and potential. However, the commander may determine through the inquiry that the report has serious irregularities or errors, such as:

- improperly designated or unqualified rating officials;
- inaccurate or untrue statements; or
- lack of objectivity or fairness by rating officials.

The commander will not pressure or force the rating officials to change their evaluations. He may not evaluate the rated NCO, either as a substitute for, or in addition to, the designated rating official's evaluation. He will not use the provisions to forward information derogatory to the rated NCO.

The inquiry must be conducted by either the commander at the time the report was rendered and is still in the command position or by a subsequent commander in the position. However, the inquiry must be forwarded to Cdr, USATAPA, ATTN: DAPC-MSE or the appropriate state adjutant general not later than 120 days after the "Thru" date of the NCO-ER (Part I, Block (h)), DA Form 2166-7.

The results of the commander's inquiry will include findings, conclusions, and recommendations in a format that can be filed with the report in the NCO's OMPF for clarification purposes. The results, therefore, will include the commander's signature, should stand alone without reference to the other documentation, and preferably be limited to one page. Reports and statements will be attached to justify the conclusion. The results of a commander's inquiry will not constitute an appeal but may be used in support of an appeal. Let me repeat that again. The results of a commander's inquiry will not constitute an appeal but may be used in support of an appeal. Having a commander's inquiry done first could very well help you get the bad report dropped. Be realistic in the assessment of whether or not to submit the appeal. Your local staff judge advocate and personnel service company are also available to advise and provide assistance in the preparation of an appeal, should you feel the need to appeal your report.

Let's look closely at the appeal process. A NCO-ER, which is inconsistent with others in an OMPF, does not mean it is inaccurate or unjust.

Appealing an evaluation report on the sole basis of a self-authored statement of disagreement will not be successful. Also, statements from rating officials claiming that they did not intend to evaluate as they did will not alone serve as the basis for altering or withdrawing an evaluation report. Careful consideration should be given before submitting an appeal of a NCO-ER in which the box markings are positive but the bullet comments are less than desirable.

If you put in an appeal, you should carefully decide what evidence is needed to support claims, whether or not such evidences are available and how to go about obtaining it. An appeal's success depends on:

- the care with which the case is prepared;
- the line of argument presented; and
- the strength of evidence presented to support it.

You need to specifically identify the entries or comments to be challenged, the perceived inaccuracy or injustice to each entry or comment, the evidence you think is necessary to prove the alleged inaccuracy or injustice, and where and how to obtain such evidence.

Third party statements form the basis of most substantive appeals. "Third parties" are persons who had knowledge of the rated NCO's duty performance during the period of the report being appealed. Statements from individuals who establish they were on hand during the contested rating period, who refute faulting remarks on the evaluation report and who served in positions from which they could observe your performance and your interactions with the rating officials, are both useful and supportive. These statements should be specific and not deal in general discussions of the appellant. For example, if you desired to challenge a rating concerning your ability to train your soldiers, it would be to your advantage to provide statements from a cross-section of individuals who can provide specific information pertaining to training your soldiers. It could be:

- the training NCO;
- platoon sergeant, platoon leader from one of the other platoons;
- one or more of your soldiers;
- your or another commander; or
- members of an evaluation team that is aware of your performance.

Such third party statements should be on letterhead, if possible, describing the author's duty relationship to you during the period of the contested NCO-ER, degree (frequency) of observation and should include the author's current address and telephone number.

If you need statements from someone that has had a permanent change of station, ETS or retired, all you have to do is check with your local worldwide locator service or call or write the one at the Department of the Army. To obtain records to verify dates, start with the 201 file. For orders and other documents, contact former organization PSC or unit level personnel offices to determine whether records are still retained.

Once you have all your statements and other information, take it to your PAC or PSC, because they will have it typed for you in the right format. Should you decide to do it yourself, check with them as to how it should be completed. Before finalizing the appeal, the entire package should be reviewed by a disinterested third party, someone that you have confidence in. Submit the finalized appeal in two complete packets directly to the appropriate address (check the PSNCO for address).

You may be told that you have up to five years to appeal an NCO-ER but do it as soon as possible because people forget, PCS, and ETS. Besides, DA feels that the sooner you do it the more important it is to you. Prompt submission is highly recommended should the need arise.

As you can see, it is not hard at all to appeal an NCO-ER; however, the best thing to do is not to get one that needs to be appealed.

Because I feel the counseling session is the key to an excellent or successful NCO-ER, let's examine some of the counseling techniques that you and your rater should use.

LISTENING AND WATCHING PRINCIPLES

Listening and watching skills involve the counselor concentrating on what the soldier says and does. Spoken words by themselves are only part of the message. The way words are arranged and spoken has meaning. The tone of the voice, the inflection, the pauses, the speed, the look on the soldier's face are all part of the total message.

Part of active listening is concentrating on what the soldier is saying. Another part is letting the soldier know the counselor is concentrating, hearing, and understanding what is said or is "getting the message." Listening skills that the counselor should consider include:

- eye contact;
- posture;
- head nod;
- facial expressions; and
- verbal behavior.

Active listening also means listening thoughtfully and deliberately to the way a soldier says things. While listening, be alert for common themes of discussion. A soldier's opening and closing statement, as well as recurring reference, may indicate the ranking of his priorities. Inconsistencies and gaps in discussion may indicate that the soldier is not discussing the real problem or is trying to hide something.

Often, a soldier who comes to the leader with a problem is not seeking help for that problem, rather he is looking for a way to get help with another, more threatening problem. Confusion and uncertainty may indicate where questions need to be asked. The elements of active listening are:

- eye contact;
- posture;

- head nod;
- facial expressions;
- verbal behavior;
- moving toward the counselor.

Maintaining eye contact helps show a sincere interest in the soldier. Occasional breaks of contact are normal and acceptable. Clock-watching, paper shuffling and excessive breaks of contact indicates a lack of interest or concern.

A relaxed and comfortable body posture helps put the soldier at ease. Being too formal or rigid makes the soldier feel uncomfortable and a too relaxed position or slouching may indicate a lack of interest.

Occasionally, nod your head as the soldier talks, it shows that you as the counselor are attentive and it also encourages the soldier to talk more.

Remain natural and relaxed. A blank look or frown may discourage the soldier from talking.

You, as the counselor, should refrain from talking too much. Let the soldier do the talking. Stay with the topic being discussed and avoid interrupting. Speaking only when necessary reinforces and stimulates the soldier. Silence can sometimes do this too. Occasional silences may indicate the soldier is free to continue talking. A long silence may distract and make the soldier uncomfortable.

While listening, you, as the counselor, must also be aware of the soldiers gestures or non-verbal behavior. These actions are part of the total message that the soldier is sending. By watching the soldier's actions, you, as the leader, can "see" the feeling behind the words.

It is important to note the difference between what the soldier is saying and doing. One common indicator to watch for is boredom, which may be displayed by drumming on the table, doodling, clicking a ballpoint pen, or resting the head in the palm of the hand. Another indicator, self-confidence, could be displayed by standing tall, leaning back with the hand behind the head, and maintaining steady eye contact.

Hate and other negative emotions may be indicated by the soldier pushing himself deeply into a chair, glaring at the counselor, and making sarcastic comments. Arms crossed or folded in front of the chest often shows defensiveness.

Frustration, another indicator, may be expressed by rubbing the eyes, pulling on an ear, taking short breaths, wringing the hand, or frequently changing total body position.

Moving towards the counselor while sitting may indicate interest, friendliness, and openness, as well as sitting on the edge of the chair with arms uncrossed and hands open.

With that information about listening and watching, let's move on and look at:

THE PROBLEM SOLVING PROCESS

There are seven basic steps of the problem solving process, which can also be used for decision-making and planning. The steps can sometimes help to structure counseling.

Here is a list of the seven steps along with some examples or guiding remarks that may fit each step, depending on the situation. The seven basic problem solving steps are:

1. Identify the problem:
- What is the cause of the problem?
- What is the biggest source of trouble?
- Tell me about what is wrong.
- Why is this a problem for you?
- How did this happen?
- I'd like to hear how you think things got this way.
- Let's list all your concerns, then we'll prioritize them.

2. Gather information:

- Let's get the facts.
- What's the background of this?
- Who is involved?
- What has been done?
- Tell me how this works.
- Describe some examples of that.

3. Develop courses of action:

- What do you want?
- How would you like things to be?
- What are some ways to do that?
- How could you get things to be the way you want?
- Let's figure out what can be done.
- What else might work?

4. Analyze and compare course of action:

- I'd like to hear about that.
- What are some problems with doing that?
- What makes that better?
- Why you are concerned with that?
- What are the disadvantages?
- What does that have to do with the problem?
- Will this get you what you want?
- How will this affect our Unit/Organization?

5. Make a decision, select a course of action:

- What solution will work best?
- Which one do you like?
- Can you describe the most likely answer?
- You need to pick a course of action.
- It's time for you to make a decision.

6. Make a plan:

- What are your next steps?
- How do you get that done?
- Now, you need a plan.
- How are you going to do that?
- Who's got to do what?

- What else must happen?
- What could go wrong?
- How can you avoid that?

7. Implement the plan:
- If you don't have any other concerns, you're ready to go.
- Now, it's time for you to act.
- Okay, get started.
- See me Monday and let me know how things turn out.
- I think you've got things figured out.
- It's all up to you now.

Remember, when you help a soldier solve a problem, you do just that—help him. Never try to do it on your own. Guide him through the problem solving process and, at the same time, be aware of the listening and watching principles.

THE COUNSELING PROCESS

Preparation is the key to a successful counseling session. Sometimes, however, planning for counseling is not possible. Leaders who know their soldiers and their duties are mentally prepared to respond to their needs.

In preparing for scheduled counseling sessions, the leader should consider the following process:

1. **Notify the soldier.** Giving the soldier advance notice involves telling him that you are going to counsel him, telling him why, where, and when the counseling is to take place, so that he can prepare for the counseling. Notifying the soldier too far in advance could make the soldier nervous and worry about the meeting.

2. **Schedule the best time ideally.** Counseling sessions should be shorter than half an hour and always less than an hour. Longer sessions become unproductive and tend to get off the subject. Complex problems need more than an hour and indicate a need for additional expertise. In deciding

when to schedule the counseling session, the counselor should select a time free from competition with other activities. Leaders should also consider what has been planned after the counseling session. If something highly important will take place, soldiers may be distracted and unable to concentrate on the counseling session.

3. **Choose a suitable place.** The place you choose for the counseling session should be free from distracting sights and sounds, it should be a place where the leader can listen to the soldier without any interruptions. Counseling is not restricted to an office; it may well be conducted in the field, motor pool, barracks, or wherever duties are being performed.

4. **Decide the right atmosphere.** A soldier at ease normally discusses matters more openly. The leader may let the soldier sit or drink a cup of coffee during the discussion. If the soldier is a smoker, try to counsel him in a place where he will be able to smoke. The counselor may not want to sit behind the desk because a desk can act as a barrier to free and open communication. The setting is important. In discipline counseling, the soldier is directed to remain standing while the leader remains seated behind a desk. This kind of atmosphere reinforces the counselor's rank, position in the chain of command, and authority as a leader.

5. **Plan the discussion.** The counselor should outline what he wants to talk about. The outline should guide the discussion but allow flexibility to react to situations that develop during the counseling. The outline should include points to discuss and the order in which to mention them. The outline is only a tool, and it should not prevent discussing the soldier's concerns. Counselors must be certain they have the necessary information, are familiar with it, and are sure of the facts. To do this!

   - Collect information and data to better understand the

soldier's ideas and attitudes.
- Summarize and organize the information to describe strengths and weaknesses or advantages and disadvantages.
- Interpret the information as it pertains to meeting established standards as well as looking for certain consistencies and patterns.
- Identify the problem from a leader's view and try to discover the cause. The leader's perspective of the problem may be different from the soldier's view, so the outline must be flexible.

The leader should also decide what approach to use during the counseling session. The more you counsel soldiers the better you will become at it, but no matter how much counseling you do you will most likely run into some pitfalls.

COUNSELING PITFALLS

A pitfall is a hidden or not easily recognized danger or difficulty. Likes, dislikes, biases, and prejudices are potential pitfalls that can interfere with the counseling relationship. Here are some common pitfalls the leader should avoid:

- **Personal Bias.** Values and ideas about the worth or importance of things, concepts and people. Personal values influence personal priorities or the desirability of different alternatives. If differences between personal values are ignored, facts can become distorted and problems further complicated.
- **Rash Judgements.** This is the tendency to evaluate a soldier on the basis of appearance or of a specific behavior trait. A halo effect may come from a significant accomplishment or from a favorable first impression. It can also result from one bad impression, from disciplinary problems, or from association with a group whose members are known to be troublemakers. After a rash judgement is made, the leader tends to ignore significant information, thus failing to develop a complete or accurate evaluation.

- **Stereotyping.** Involves judging soldiers on presumed group physical or behavioral characteristics. Evaluations should be made only on a soldier's demonstrated behavior or on his demonstrated ability and not on presumed physical, racial, or other characteristics. Leaders should not stereotype soldiers or let stereotyping affect an evaluation or recommendation.
- **Loss of Emotional Control.** The advantage of self-control to a leader applies to him in his role as counselor. If the counselor can control his emotions, most likely the soldier will too. If the counselor loses control of the session, little will be accomplished. Differences of opinion are acceptable, but arguing, debating, or having a heated discussion is not. While there may be disagreement with a philosophy or attitude, it should not influence the evaluation of the situation.
- **Inflexible Methods.** Soldiers will vary according to their individual personalities, experiences, education, problems, situations, and surroundings. The same counseling approach or technique will not be effective for all. Leaders must know each soldier's individuality and adapt their approach accordingly.
- **Amateur Character Analysis.** Leaders must recognize and accept their limitations in counseling soldiers. The temptation to become an amateur psychologist or psychiatrist must be avoided. Leaders should not try to determine or to change deep-seated personality disorders that certain actions of the soldier might indicate. Counselors must be able to identify those situations which are clearly beyond their capabilities and refer the soldier to the appropriate support agency for help.
- **Improper Follow-up.** Proper and complete follow-up is important to retain and strengthen the soldier's confidence in his leader. Unkept promises to one soldier will cause the loss of confidence and respect of other soldiers. Follow-up is especially important when a soldier is referred to an agency for assistance. Because of his referral, the soldier may feel that the leader no longer cares.
- **Reluctance to Counsel.** Young, inexperienced leaders

> often hesitate to counsel subordinates on areas for improvement. Some junior leaders may want to avoid the unpleasant duty of discussing shortcomings for fear of becoming unpopular. Others may be reluctant to counsel soldiers who have been in the unit or service longer than they have but, without the counseling effort, problems get worse. These young leaders must realize that they have a responsibility to counsel soldiers fairly and objectively and that their seniors will assist them in developing their counseling skills.

As a NCO, sooner or later, you will have to be the rater for another NCO. As time moves on, you could very well be a senior rater. Even if you never have to rate another NCO, you most likely will have to inform your soldiers about their duty performance, which is the main reason for all the information about counseling.

PERFORMANCE COUNSELING OF INDIVIDUALS

All leaders need to do performance counseling of individuals. Performance counseling informs the individual about his job and the expected performance standards and provides feedback on actual performance. A soldier's performance includes appearance, conduct, mission accomplishment, and the way duties are carried out.

The purpose of counseling may be to help a soldier maintain or improve a satisfactory level of performance or improve performance that is below standards.

Good leaders issue clear guidance and then give honest feedback to let soldiers know how they have performed. Honest feedback is essential for motivating soldiers and controlling a unit's performance. The leader first observes the soldier's performance of duty, his ability to complete an assignment, and his approach to accomplishing a mission. The leader then tells the soldier where he stands. Those things that have been done well or that show improvement must be praised. The contribution that the soldier's performance has made to the unit should be noted.

Performance counseling needs to be done continuously as part of the leader's role as a teacher and as a coach. Performance counseling must be a teaching process with continuous growth and development as its object. The leader must know the soldier's character, preferences, ambitions, qualifications, and potential.

Motivation results from learning and it is greatly influenced by personal value. Those conditions under the leader's control that stimulate learning and motivation includes:

- accurate evaluation of performance;
- rapport between leaders and soldiers;
- clear and understandable communication between leaders and subordinates;
- mutual agreement concerning performance areas where improvement is required;
- specific actions for improving performance;
- feedback on progress; and
- expectation of success.

Performance counseling begins with evaluating the soldier's performance or action. It should be restricted to appraising and discussing observed actions and demonstrated behavior rather than diagnosing character or suspected attitudes.

One way to structure performance counseling is to use evaluation report forms. A monthly or quarterly review of the soldier's actions can be done using these forms as outlines to discuss specific duties and performance objectives. This ensures that soldiers receive periodic feedback on all aspects of performance that will be formally evaluated. Some soldiers require that the counselor be directive and list item by item what must be done to improve. Other soldiers, with some non-directive guidance, can figure out what to do.

Determining ways to improve is based on the leader's first evaluation of soldier's performance. Specific actions must be viewed to figure out why a soldier is below standard in a given area. It may be that the soldier does not know how, does not want to do something, or there is something that prevents proper performance. For each reason, steps needed to improve performance are different. If the soldier does not know how to do something, he needs to take steps to practice and learn.

By discussing specific actions, the leader will be far more effective in helping soldiers improve their performance. That's why all leaders must do their very best when conducting a conseling session.

CONDUCTING A COUNSELING SESSION

A counseling session can be divided into three phases: opening the session, discussion and closing the session.

Opening the Session. How the session is opened largely determines its effectiveness. Since nervousness and tension are easily detected, the counselor must create an atmosphere that will not disturb the soldier. When using the directive approach especially, the leader must appear confident and in control of the situation. When using the non-directive approach, the soldier must feel relaxed and free to speak openly. The leader's first actions and remarks help establish the desired atmosphere.

The first objective is to establish rapport with the soldier and reduce any uncertainty. The second objective is to explain the reason and to outline the conduct of the counseling session, that should establish the structure, set general time limits and discuss the degree of confidentiality at the start of the session.

Discussion. The leader must ensure that effective two-way communication is taking place during this phase. Both parties must have a clear understanding. In using the problem-solving process, the leader gathers information and then causes the soldier to define the problem, develop courses of action, select the best solution and implement it. Career and other types of counseling will require different steps. If there is a misunderstanding, the leader must clear it up and, if the problem is beyond his ability, he should refer the soldier to the appropriate support agency.

Closing the Session. In closing the counseling session, the leader must summarize what has been discussed. The counselor must ensure that both parties understand what each is expected to do. This can be done by having the soldier review what he is going to do and expects the leader to do. Any additional questions can be answered but the closing is not the time to bring up new information. Any future meeting should be scheduled before the soldier is dismissed. The leader's duties have not been fully performed when the counseling session ends. The leader must either act on or follow-up on what was discussed.

If you started reading from the beginning of this book, you should know about the rating chain qualifications and responsibilities, the checklist and working copy, how to fill out the DA Form 2166-7, why bullet comments are important and how to write them, types of NCO-ER's and how to counsel your soldier. With that information, you should be able to write the type of NCO-ER needed for your NCO's. More NCO's are being forced out of the Army because of bad NCO-ER's than any other reason. Before long, there will be another way to force out NCO's and, because of that, I feel it would not be fair if I didn't say something about how it can affect your promotions, assignments, school selections, and retention.

If it's the last thing you do, please be sure to read the rest of this book. It is just as important as the first part.

THE SELF DEVELOPMENT TEST

The Self Development Test was not developed to replace the old Skill Qualification Test because not only does it contain questions about the NCO's MOS but also questions about leadership and training. The SDT will be taken by all Sergeants, E-5 to E-7. Because the SDT has the same effect on your career as the NCO-ER, it would stand to reason that failing the SDT could also cause you to be forced out of the Army or selected for QMP. Looking at it another way, if you should get an Excellent rating for training on your NCO-ER but fail the training part of the Self Development Test, how do you think DA will look at it? When the NCO takes the SDT, he will be asked questions about his MOS (60 percent), Training (20 percent), and

Leadership (20 percent). To study for the test, the NCO will need to study:

- FM 22-100 Military Leadership;
- FM 22-101 Leadership Counseling;
- FM 22-102 Soldier Team Development;
- FM 25-101 Battle Focused Training;
- all your MOS Manuals.

Due to the fact that 60 percent of the test will be about the MOS, I feel most NCO's will pass the test. The SDT only calls for a 70 percent overall to pass the test. However, the NCO will have to read the books and make sure he understands what he has read. To do this, he must know how to study for an SDT.

HOW TO STUDY FOR THE SDT

To study for the SDT, you must study the same way you study for any other test. You must:

- examine each chapter;
- read for the goals and situations;
- ask questions as you review the book;
- organize your study and textbook notes;
- underline notes; and
- review regularly.

Examine each chapter. Before you read the book, you will learn the scope of the contents, how the topics are arranged, and the writer's purpose and point of view. The chapter title will contain the chapter's main idea, as well as the opening sentence or two. If there is a chapter objective, it will tell you specifically what information you will gain from the chapter. The chapter summary or review will contain all the major points covered in the chapter, it will tell you which are the important ideas on which to concentrate as you read the chapter. The major heading and subheading will provide the outline of the chapter.

Read for the goals and situation. Concentrate on what you are reading, pay close attention to the headings and subheadings. They will indicate the relative importance of each topic. Find the main ideas in each chapter or section. Most writer's develop a topic sentence and/or paragraph, substantiate it, and draw conclusions. Recite the main ideas to yourself after finishing a page and check to see if you were correct. Do the same for the major points after reading each chapter or section. The first and last sections of most reading material will usually state the most important facts and information about the topic so examine them carefully. Find the main ideas and then the important details that support them.

Ask questions as you review the book. When you start examining a book, ask yourself questions as if you are talking to the writer of the book. Ask, "What does the chapter title mean? The heading and subheading?" Ask, "Why the writer developed the material in the order presented? Would you get the same conclusions if the material were presented in a different arrangement?" Take this book for instance. Do you think you would get the same benefits if Chapter One was last? Ask, "How does the writer support the main heading and subheading? Does he address you as someone sympathetic to his position or as a potential follower? What are his qualifications and how is he associated with the subject?"

Organize your study and textbook notes. Develop your own note-taking techniques. Many people use only one side of the paper for their notes, leaving a two or three inch margin on the left side of the same page for writing key words and labelling. If you have classroom notes, you can put your class notes on the right hand page of your notebook and transfer your text notes to the appropriate left-hand facing page. This way you can easily review all the information gained from your class and text reading. When taking notes, make sure you understand what you are writing. A good dictionary and thesaurus are a must for effective note taking. If the book has a glossary, refer to it because it will define words that are sometimes unique to the subject at hand.

Underline notes. Marking your books increases your understanding of the material for the present and for future reference. The process of selecting and marking requires you to find the main ideas. Later, when you review the book for the test, you will find that your marking and understanding enable you to grasp the essential points without having to read the entire paragraphs and chapters again. Underlining key words and sentences will make those items stand out in your mind.

Summaries enable you to write a brief summation of a section in your own words. You could use a bracket to enclose several consecutive lines that are important. Rather than underlining all of them, you could box or circle key terms. Develop your own system, but read before you mark. Read a few paragraphs or sections and then go back over the material and underline those topics and/or words that you feel are important ones. Underline only those points that you feel are clearly essential.

Review regularly. Reviewing is an essential part of retention. Obviously, you will reap the most benefit from reviewing your own notes in your own book. Review your notes shortly after you have written them and continue to review them periodically. Spend a few minutes going over your earlier notes before beginning a new reading assignment. Constant review throughout the chapters and book will greatly reduce the time you'll need to spend preparing for the test, and make that time less "cram" filled and far more relaxing.

Remember the more you review this book and write NCO-ER's the better you will become at doing it. The Army is always trying to find better ways of doing things. While the form for doing the NCO-ER may change, I feel the bullet comments are here to stay.

Use this book and let your friend use it or tell them where they will be able to buy one.

Whatever you do, take charge of your career by taking charge of your NCO-ER.

MY PERSONAL WORD LIST

MY PERSONAL BULLET COMMENTS

CAREER RESOURCES

Contact Impact Publications to receive a free copy of their latest comprehensive and annotated catalog of career resources (books, subscriptions, training programs, videos, audiocassettes, computer software, and CD-ROM).

The following career resources are available directly from Impact Publications. Complete the following form or list the titles, include postage (see formula at the end), enclose payment, and send your order to:

IMPACT PUBLICATIONS
9104-N Manassas Drive
Manassas Park, VA 22111
Tel. 703/361-7300
FAX 703/335-9486

Orders from individuals must be prepaid by check, moneyorder, Visa or MasterCard number. We accept telephone and FAX orders with a Visa or MasterCard number.

Qty.	TITLES	Price	TOTAL
	MILITARY		
___	America's Top Military Careers	$19.95	______
___	Army Officer's Guide	$19.95	______
___	Beyond the Uniform	$12.95	______
___	Civilian Career Guide	$12.95	______
___	Complete Guide to the NCO-ER	$13.95	______
___	Enlisted Soldier's Guide	$12.95	______
___	From Army Green to Corporate Gray	$13.95	______
___	Guide to Military Installations	$17.95	______
___	How the Military Will Help You Pay For College	$9.95	______
___	Job Search: Marketing Your Military Experience	$14.95	______
___	NCO Guide	$17.95	______
___	Re-Entry	$13.95	______
___	Retiring From the Military	$22.95	______
___	Self-Development Test Study Guide	$14.95	______
___	Today's Military Wife	$14.95	______

	Title	Price	
___	Up or Out: How to Get Promoted as the Army Draws Down	$13.95	______
___	Veteran's Survival Guide to Good Jobs	$12.95	______

JOB SEARCH STRATEGIES AND TACTICS

	Title	Price	
___	Change Your Job, Change Your Life	$14.95	______
___	Complete Job Finder's Guide to the 90s	$13.95	______
___	Dynamite Tele-Search	$10.95	______
___	Electronic Job Search Revolution	$12.95	______
___	Five Secrets to Finding a Job	$12.95	______
___	How to Get Interviews From Classified Job Ads	$14.95	______
___	Professional's Job Finder	$18.95	______

BEST JOBS AND EMPLOYERS FOR THE 90s

	Title	Price	
___	100 Best Companies to Work for in America	$27.95	______
___	American Almanac of Jobs and Salaries	$17.00	______
___	Best Jobs for the 1990s and Into the 21st Century	$12.95	______
___	Job Seeker's Guide to 1000 Top Employers	$22.95	______
___	Jobs Rated Almanac	$15.95	______

KEY DIRECTORIES

	Title	Price	
___	Career Training Sourcebook	$24.95	______
___	Dictionary of Occupational Titles	$39.95	______
___	Directory of Executive Recruiters (annual)	$44.95	______
___	Moving and Relocation Directory	$149.00	______
___	National Directory of Addresses & Telephone Numbers	$129.95	______
___	National Trade and Professional Associations	$79.95	______
___	Occupational Outlook Handbook	$22.95	______
___	Places Rated Almanac	$21.95	______

CITY AND STATE JOB FINDERS (Bob Adams JobBanks)

	Title	Price	
___	Atlanta	$15.95	______
___	Boston	$15.95	______
___	Chicago	$15.95	______
___	Dallas/Fort Worth	$15.95	______
___	Denver	$15.95	______
___	Florida	$15.95	______
___	Los Angeles	$15.95	______
___	Minneapolis	$15.95	______
___	New York	$15.95	______
___	Seattle	$15.95	______
___	Washington, DC	$15.95	______

ALTERNATIVE JOBS AND CAREERS

	Title	Price	
___	Business and Finance Career Directory	$17.95	______
___	But What If I Don't Want to Go to College?	$10.95	______
___	Careers in Computers	$16.95	______
___	Careers in Health Care	$16.95	______
___	Environmental Career Guide	$14.95	______
___	Travel and Hospitality Career Directory	$17.95	______

INTERNATIONAL, OVERSEAS, AND TRAVEL JOBS

	Title	Price	
___	Almanac of International Jobs and Careers	$14.95	______
___	Complete Guide to International Jobs & Careers	$13.95	______
___	Guide to Careers in World Affairs	$14.95	______
___	How to Get a Job in Europe	$17.95	______
___	Jobs for People Who Love Travel	$12.95	______
___	Jobs in Russia and the Newly Independent States	$15.95	______

PUBLIC-ORIENTED CAREERS

	Title	Price	
___	Almanac of American Government Jobs and Careers	$14.95	______
___	Complete Guide to Public Employment	$19.95	______
___	Federal Jobs in Law Enforcement	$15.95	______
___	Find a Federal Job Fast!	$9.95	______
___	Government Job Finder	$14.95	______
___	The Right SF 171 Writer	$19.95	______

JOB LISTINGS & VACANCY ANNOUNCEMENTS

	Title	Price	
___	Federal Career Opportunities (6 biweekly issues)	$38.00	______
___	International Employment Gazette (6 biweekly issues)	$35.00	______
___	The Search Bulletin (6 issues)	$97.00	______

SKILLS, TESTING, SELF-ASSESSMENT

	Title	Price	
___	Discover the Best Jobs for You	$11.95	______
___	Do What You Love, the Money Will Follow	$10.95	______
___	What Color Is Your Parachute?	$14.95	______

RESUMES, LETTERS, NETWORKING, DRESS

	Title	Price	
___	Dynamite Cover Letters	$9.95	______
___	Dynamite Resumes	$9.95	______
___	Electronic Resume Revolution	$12.95	______
___	Electronic Resumes for the New Job Market	$11.95	______
___	Great Connections	$11.95	______
___	High Impact Resumes and Letters	$12.95	______
___	Job Search Letters That Get Results	$12.95	______
___	New Network Your Way to Job and Career Success	$12.95	______
___	Red Socks Don't Work!	$14.95	______

INTERVIEWS & SALARY NEGOTIATIONS

	Title	Price	
___	60 Seconds and You're Hired!	$9.95	______
___	Dynamite Answers to Interview Questions	$9.95	______
___	Dynamite Salary Negotiation	$12.95	______
___	Interview for Success	$11.95	______

WOMEN AND SPOUSES

	Title	Price	
___	Doing It All Isn't Everything	$19.95	______
___	New Relocating Spouse's Guide to Employment	$14.95	______
___	Resumes for Re-Entry: A Handbook for Women	$10.95	______
___	Survival Guide for Women	$16.95	______

MINORITIES AND DISABLED

___	Best Companies for Minorities	$12.00	______
___	Directory of Special Programs for Minority Group Members	$31.95	______
___	Job Strategies for People With Disabilities	$14.95	______
___	Minority Organizations	$49.95	______
___	Work, Sister, Work	$19.95	______

ENTREPRENEURSHIP AND SELF-EMPLOYMENT

___	101 Best Businesses to Start	$15.00	______
___	Best Home-Based Businesses for the 90s	$10.95	______
___	Entrepreneur's Guide to Starting a Successful Business	$16.95	______
___	Have You Got What It Takes?	$12.95	______

SUBTOTAL ______

Virginia residents add 4½% sales tax ______

POSTAGE/HANDLING ($3.00 for first title and $1.00 for each additional book) $3.00

Number of additional titles x $1.00 ---------- ______

TOTAL ENCLOSED ---------------- ______

NAME ____________________

ADDRESS ____________________

____

____

[ ] I enclose check/moneyorder for $ ________ made payable to IMPACT PUBLICATIONS.

[ ] Please charge $ ________ to my credit card:

Card # ____________________

Expiration date: _____/_____

Signature ____________________